87 Advances in Polymer Science

Conformational Motion and Disorder in Low and High Molecular Mass Crystals

By B. Wunderlich, M. Möller,
J. Grebowicz, H. Baur

With 42 Figures and 11 Tables

Springer-Verlag Berlin Heidelberg GmbH

Bernhard Wunderlich

Department of Chemisuy. The University of Tennessee, Knoxville, TN 37996-1600, USA

Martin Möller

Institut für Makromolekulare Chemie, Universität Freiburg, D-7800 Freiburg, FRG

Janusz Grebowicz

Dept. Polymer Science and Engineering, University of Massachusetts, Amherst, MA 01003, USA

Herbert Baur

Kunststofflaboratorium, BASF AG, D-6700 Ludwigshafen, FRG

ISBN 978-3-662-15115-0 ISBN 978-3-540-38867-8 (eBook)
DOI 10.1007/978-3-540-38867-8

Library of Congress Catalog Card Number 61-642

Originally published by Springer-Verlag Berlin Heidelberg New York in 1988
Softcover reprint of the hardcover 1st edition 1988

2152/3020-543210

Editors

Table of Contents

The broad field of conformational motion and disorder in crystals is reviewed with attention paid to the distinction of the recently defined condis crystals from the well-known mesophases of liquid crystals and plastic crystals. Structure, thermodynamics, motion and transitions of a number of small and large molecules are discussed. The cooperative nature of the defect equilibrium is analyzed. Of special interest are the borderlines between high viscosity liquid crystals and condis crystals and between plastic crystals and condis crystals. The complications posed by pseudorotation, jumping between symmetry-related states and freezing into non-equilibrium states are illuminated.

Erratum: Vol. 86, p. 143

Editor of the contribution is Prof. Dr. H. Höcker, Aachen

1 Introduction

1.1 Definition of Condis Crystals and an Approximate Empirical Transition Entropy Description

In a recent review of thermotropic mesophases and their transitions [1] it was suggested that crystals with dynamic conformational disorder do not fit into the standard classification of mesophases as either plastic crystals [2] (with dynamic orientational disorder and long-range positional order) or liquid crystal [3] (with positional dynamic disorder and some long-range orientational order). At that time it was proposed that a distinct third type of mesophase exists, that of the condis crystal (with dynamic *con*formational *dis*order and long-range positional and orientational order). Fig. 1.1 illustrates schematically the various condensed phases and their interrelations. The three classical condensed phases: glass, crystal and melt, are outlined doubly. Glasses and crystals are considered solid, the melt liquid. Condis crystal, plastic crystal, and liquid crystal are increasingly less solid (more liquid). The arrows on the right-hand side indicate the possible transitions between the phases that are often first order transitions, i.e. transitions that involve a discontinuity in the first derivative of the Gibbs energy function, the entropy [4]. When starting with the crystal, the entropy increases going to successively lower states drawn in Fig. 1.1.

The overall entropy of fusion ΔS_f, of a crystal can be approximately separated into the three contributions, arising from positional disordering, ΔS_{pos}, orientational disordering, ΔS_{orient}, and conformational disordering, ΔS_{conf}, so that

$$\Delta S_f = \Delta S_{pos} + \Delta S_{orient} + \Delta S_{conf} . \tag{1.1}$$

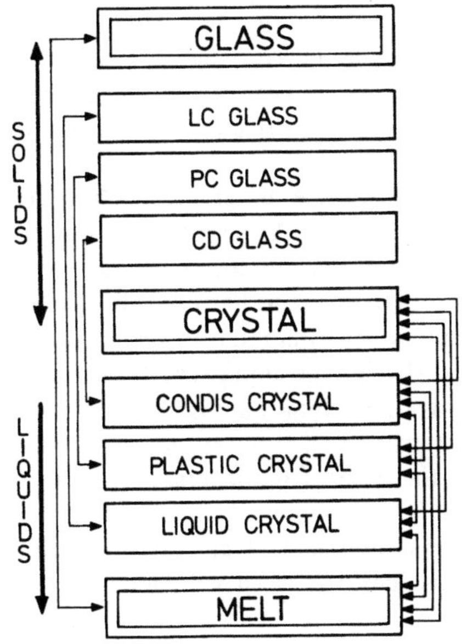

Fig. 1.1. Schematic diagram of the relationship between the three limiting phases (double outline) and the six mesophases. Adapted from Ref. [1]

It is possible to make a first judgment of a phase classification based on the thermal analysis of the first order transitions. Such analysis is helped by the empirical "rules" that ΔS_{pos} is generally 7–14 J/(K mol) (Richards' rule) [5]. $\Delta S_{o\ ient}$ is usually 20–50 J/(K mol) (based on Walden's rule) [6], and ΔS_{conf} is about 7–12 J/(K mol) [7]. While ΔS_{pos} and ΔS_{orient} are referred to the whole molecule, i.e. they do not change with size; ΔS_{conf} is referred to a flexible bond (or a rigid bead), i.e. for a molecule, ΔS_{conf} is size-dependent. The rather broad ranges of values indicate the approximate nature of this description. For a more precise analysis, effects of disorder, volume change, changes in vibrational frequency etc. would have to be detailed. Nevertheless, the empirical rules permit estimations of the various transition entropies corresponding to the arrows of the right-hand side of Fig. 1.1. Particularly well obeyed is a transition entropy of 7–14 J/(K mol) for the transition from the plastic crystal to the melt [2]. The analogous transitions from the liquid crystalline phase to the melt are always much less than ΔS_{orient}, indicative of the rather high degree of orientational disorder and mobility in liquid crystals [1]. Condis crystals, finally, are expected to show variable entropies of transition from the crystal to the condis-crystal and from the condis-crystal to the isotropic melt, depending on the type and concentration of conformational disorder.

On cooling, the mesophases do not always follow the equilibrium path. If the ordering to the crystal can be by-passed, the liquid-like cooperative motion of the mesophases freezes at a glass transition, as is indicated on the left side of Fig. 1.1. At these glass transitions there is no change in order and, as a result, no change in entropy. Since the liquid-like motion changes at the transition to practically only vibrational motion, there is a drop in heat capacity on vitrification. For liquid crystals, the existence of the glassy state was already implied in the work of Vorlaender [8]. Their drop in heat capacity is practically identical to those for liquids [1]. For plastic crystals, the glassy state was first recognized by Adachi, Suga, and Seki [9]. The order of magnitude of the change in heat capacity at the glass transition of plastic crystals is also not much different from that of liquids. For the typical glass transition of a liquid it was found that the change in heat capacity is approximately 10–20 J/(K mol) when calculated per mole of mobile parts (beads) within the molecule [10]. Since condis crystals show usually intermediate heat capacities between the solid and liquid states, their change in heat capacity on vitrification is expected to be variable also.

In this discussion at attempt will be made to describe in greater detail the structure and motion for a larger number of condis crystals. A special effort will be made to point-out the differences between condis crystals on the one hand, and liquid and plastic crystals on the other. It seems reasonable, and has been illustrated on several examples, that molecules with dynamic, conformational disorder in the liquid state show such conformational disorder also in the liquid crystalline and plastic crystalline states [11]. The major need in distinguishing condis crystals from other mesophases is thus the identification of translational motion and positional disorder of the molecular centers of gravity in the case of liquid crystals, and of molecular rotation in the case of plastic crystals.

Polymorphism is known for mesophases as well as for crystals. It can be considered a subdivision of the scheme of Fig. 1.1. In most cases it is easy to identify the polymorphs as one of the four crystal phases, altough particularly some of the highly ordered smectic liquid crystals are difficult to distinguish from crystals or condis crystals as long as a detailed knowledge of structure, motion and thermodynamics is not available. Little can be offered in this review to resolve such problems of polymorphism.

1.2 Justification of the Empirical Transition Entropy Description

To find an explanation for Richards' rule discussed in Sect. 1.1, one may postulate that the liquid state possesses a short-range "quasi-crystalline" structure [12]. As the temperature increases, this structure changes gradually, to become "quasi-gaseous" close to the critical temperature [12]. The melting process may be, based on this model, broken into three major steps. First, the crystal lattice is expanded to give the average separation of the motifs as it exists in the melt. Next, the motifs are dynamically disordered. Last, defects are introduced into the quasi-crystalline melt. Starting from a cubic close-packed crystal Lennard-Jones and Devonshire computed the three melting stages assuming interstitial atoms as defects [13]. For argon an entropy of fusion of 1.70 R, or 14.1 J/(K mol), was computed that matches the experimental equilibrium value of 14.0 J/(K mol) at 83.8 K. The entropy increase in going from fixed motifs in the small volume element of the crystal to the total volume of the melt (communal entropy) was earlier found to be R, or 8.31 J/(K mol) [14]. The lower and upper limit of Richards' rule seem thus reasonable.

In analogy to the estimates of ΔS_{pos}, it is possible to suggest that the empirically observed values for ΔS_{conf} can be linked to the commonly used "rotational isomeric states" model of linear macromolecules [15–17]. Assuming two to four rotational isomers to become dynamically accessible on a disordering transition and 75% of the total entropy change to be conformational, one expects an entropy gain of 7.7 to 15,3 J/(K mol), the correct size range of the experimental values. On the basis of atomistic calculations, a detailed rotator model was analyzed for the condis phase of flexible linear macromolecules using a cooperative statistical treatment [18]. Both, the transition from the crystal to the condis crystal and from the condis crystal to the melt could be described quantitatively for polyethylene and for trans-1,4-polybutadiene as a function of pressure. The overall melting of macromolecules was treated as an, at least, two-step transition: a conformational transition and the cooperative formation of dislocations, the latter giving rise to only minor changes in conformation [19,20]. For the condis states of polyethylene and possibly also trans-1,4-polybutadiene the two steps occur largely separated at the two equilibrium transition temperatures T_d and T_i (disordering and isotropization temperatures, respectively).

The orientational entropy of fusion $\Delta S_{orient} = 20$–50 J/(K mol) discussed in Sect. 1.1 is based on Walden's empirical rule of entropies of irregular, rigid molecules [6]. Subtraction of the appropriate ΔS_{pos} leads to the suggested values. The separation of ΔS_{pos} and ΔS_{orient} is well documented by the plastic crystals that often show ΔS_{pos} as the isotropization transition [2].

1.3 Motion in the Condensed Phase

The major motion type for the crystalline and glassy solid states is vibration. For linear macromolecules one finds large deviations from the classical Einstein [21] or Debye [22]

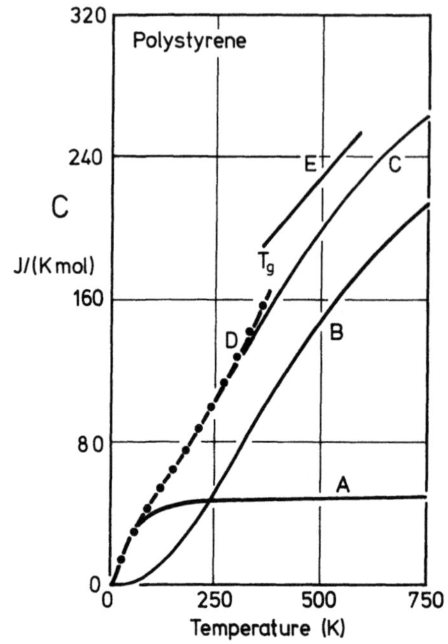

Fig. 1.2. Computed and experimental heat capacities of polystyrene. *Glassy solid state* A: skeletal contribution (Tarasov function, θ_1 = 284 K, θ_3 = 48 K, N_s = 6); B: group contributions (N_g = 42); C: total, computed heat capacity at constant volume; D: computed heat capacity at constant pressure and experimental points. *Liquid state* E: experimental data between the glass transition, T_g = 373 K, and 600 K (heat capacity at constant pressure)

treatments of the heat capacity. Also, an early attempt to approximate the heat capacities with a combination of a 1-dimensional and 3-dimensional continuum to account for the anisotropy of the chain molecules [23] was by itself not applicable over a wide temperature range. The method that was finally successful [24,25] involves an approximation of the frequency spectrum by separation of the high-frequency group vibrations from the skeletal chain-vibrations. The group vibrations are then accessible from normal-mode calculations on isolated chains and are even transferable for polymers of similar chemical structure. The skeletal vibrations can then be approximated using the two-parameter Tarasov approach (θ_1 and θ_3) [23]. Agreement between computation and experiment has been tested for almost 100 polymers and is usually better than $\pm 5\%$ over the whole temperature range, starting from about 10 K [26,27]. Fig. 1.2 illustrates the computed heat capacities of polystyrene, a polymer with six skeletal vibrations and 48 group vibrations per repeating unit. Within the small scale of the drawing there is hardly any difference between experiment and computation [28].

Also plotted in Fig. 1.2 is the experimental heat capacity of the liquid (at constant pressure) [28]. In simple cases, such as polyethylene, the heat capacity of the liquid state could be understood by introducing a heat capacity contribution for the excess volume (hole theory) and by assuming that the torsional skeletal vibration can be treated as a hindered rotator [29]. A more general treatment makes use of a separation of the partition function into the vibrational part (approximated for heat capacity by the spectrum of the solid), a conformational part (approximated by the usual conformational statistics) [15–17], and an external or configurational part. The latter was shown to be almost completely represented by the difference of heat capacity at constant pressure and volume [30]. The heat capacity of all polymers is thus understood well within the precision-range of the experiment.

Adding the measured heats of transition to the integrated heat capacities it is then possible to compute not only enthalpies, but also absolute entropies and free enthalpies. The change in entropy with temperature and state of matter is illustrated in Fig. 1.3 for the example of 1,4-polybutadienes [31]. The corresponding heat capacity and transition plot is shown in Fig. 4.10, below. The vibrational contributions to the entropies of the two isomers differ only little ($\theta_1 = 589$ K for the cis-isomer and 599 K for the trans-isomer). The sum of the entropies of the two stages of fusion of the trans-isomer are close to the single-step entropy of fusion of the cis-isomer. Typical is also the much larger accumulated vibrational entropy when compared to the transition entropy.

In the course of analysis of motion types in crystals and mesophases, it became clear that large scale motion is also possible in molecular crystals. Conformational, rotational, and even diffusional motion can start considerably below the melting or disordering transition temperatures. The major tools for the determination of this motion, commonly connected only with the liquid or mesophase, are the various forms of solid state NMR [32]. In order not to lead to dynamic disorder, these larger scale motions must be jump-motions between positions of identical symmetry within the crystal. Since the intermediate in a jump-motion is present only during a negligible fraction of the total time, it can hardly be determined by most structure-analysis techniques such as X-ray diffraction and Raman or infrared spectroscopy. Similarly, there are hardly any effects thermodynamically as long as the order (entropy) sees little change [33]. The changes in heat capacity when a torsional oscillator changes to a hindered rotator, for example, are gradual and occur usually over a temperature

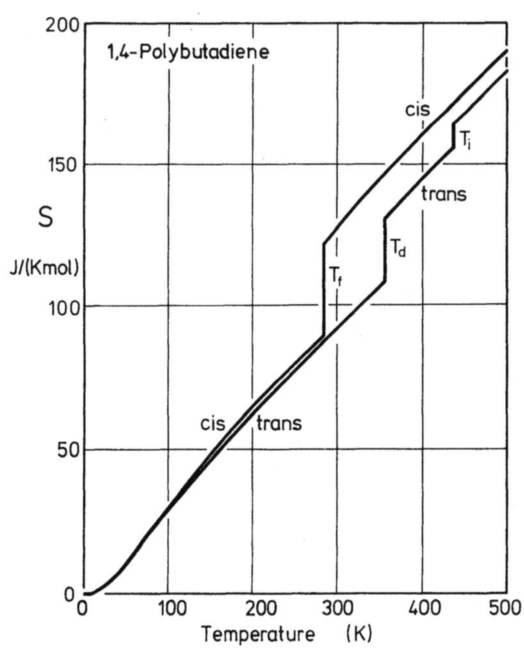

Fig. 1.3. Entropy of trans-1,4-polybutadiene that shows a condis phase between T_d and T_c and of cis-1,4-polybutadiene that does not show a condis phase, but melts in one step at T_f

range of several hundred kelvins so that its initial jumping from minimum to minimum in potential energy is difficult to distinguish from a pure oscillator [25,29]. A larger number of such jump-motions are discussed in Sections 3–5 on the basis of NMR data (see Figs. 3.1, 3.2 and 5.2, for example). Ultimately this motion leads always to disordering of the crystal to mesophases or the melt.

Another large amplitude molecular motion type, first analyzed in the gas phase of pentane [34] is the pseudorotation [35]. It is an internal motion by changing mainly torsional angles of the nuclei which interconverts conformers that are identical except for the numbering of the atoms. To visualize the motion, one may start for cyclopentane, for example, with the molecule in the planar C_5 symmetry. This symmetry represents a local maximum in potential energy. There are then 10 puckered conformers of C_s symmetry and 10 twisted conformers of C_2 symmetry, both types having almost the same low potential energy. Interchange within a group of isomeric conformers appears as a rotation about the axis normal to the plane of the molecule due to the indistinguishable nature of the CH_2-groups, has, however, no momentum about this axis. The motion actually consists of displacements parallel to the pseudorotation axis of corresponding groups by rotational motion about bonds.

In the crystal, pseudorotation is strongly hindered at sufficiently low temperature because of the volume restrictions resulting from the closely packed structure. On changing to the plastic crystal phase this restriction is removed, and pseudorotation is possible as in the liquid state (see Sect. 3.1.1). Larger cycloalkanes, described in Sect. 3.1.2, do not display a plastic crystalline state because of their larger deviation from a spherical shape. Still, these molecules show fast conformational interchange below any disordering transition temperature. This conformational interchange must occur largely within the confines of the corrugated, tubular volume available for the molecule within the crystal. Motion restricted to such a geometry can only be a concerted rotation and pseudorotation. The molecule must in this case reptate [36] along a corrugated circular path. To distinguish this type of motion from true rotation and pseudorotation we call it "rotoreptation". As in true rotations one expects no entropy gain in rotoreptation if the motion consists of jumps between positions of identical symmetry. In this case one would also not detect structure changes in the crystal. More details about entropy changes observed as soon as changes in the crystalline order take place are described in Sects. 3.1.2 to 4.

1.4 Summary and Conclusions

In this Introduction an extended subdivision of condensed phases is offered in Fig. 1.1. The dynamic disorder types may be used as basis for operational definitions of the crystal phases, and the corresponding static disorder, as basis for the glass phases. Empirical and semiempirical rules of entropy and changes in heat capacity permit to set up macroscopic criteria for the distinction of the phases. More direct measurements on a microscopic, atomic level must involve both, identification of the proper type of large-amplitude motion *and* disorder for the four crystal phases. The corresponding glasses can then be identified by the freezing of the large-amplitude motion. It was a surprise to the authors and possibly also to many readers that conformational, rotational and even diffusional motion can occur in a jump-like fashion and have then only negligible influence on the disorder in a crystal. Both, motion and disorder must thus be proven for a given mesophase.

In the following Sections it will first be discussed that there is a good chance that, particularly for different condis crystals, a full spectrum of increasingly more cooperative transitions may be possible. Conformational disorder that can be introduced without much cooperation from neighboring groups appears gradually, without first order transition. Such condis states are difficult to identify by thermal means, and microscopic techniques must be used for the identification of dynamic disorder. In the major body of the review, Sections 3–6, many condis crystal examples of small molecules and macromolecules have been treated in sufficient detail to allow clear identification of the mesophases, or at least permit an "educated guess" of the phase-nature.

2 Thermodynamics of the Conformational Isomerism

A system of molecules with various conformational isomers is a mixture from the point of view of phenomenological thermodynamics. The composition cannot, however, be specified because the components of the mixture, i.e. the physically distinguishable conformational isomers, are continuously interconverting in a kind of quasi-chemical reaction that depends on the thermal agitation. The conformational isomerism gives the system internal degrees of freedom that are described phenomenologically by so-called internal variables. The mean mass concentrations or mole fractions of the conformational isomers can, for example, be chosen as internal variables. If the conformational degrees of freedom are coupled, linear combinations of these variables are determinant. Their composition may depend on external variables such as temperature, pressure, mechanical stress, etc. [37].

Values for the internal variables in thermodynamic, internal equilibrium are generally uniquely defined by the values for the external variables. For instance, in a simple, thermomechanical system (i.e. one that reacts mechanically solely volume-elastically) the equilibrium concentrations of the conformational isomers are uniquely described by temperature and pressure. In this case the conformational isomerism is not explicitly perceptible, but causes only overall effects, for example in the system's enthalpy or entropy. Drastic macroscopic effects may, however, occur when the relationship between internal and external variables is not single-valued. Then the response-functions of the system diverge or show discontinuities. The system undergoes a thermodynamic transformation. The best-known example of such a transformation based on conformational isomerism is the helix-coil transition displayed by some polymers in solution. An example in the solid state is the crystal-to-condis crystal transition discussed in this paper. The conditions under which such transformations occur are dealt with in more detail in Sect. 2.2.

In non-equilibrium the internal variables are independent of the external variables. In this case conformational isomerism occurs also explicitly if it is thermodynamically relevant. Significant is then the experimental observation that conformational isomers may attain internal equilibrium relatively slowly. This entails that conformational isomerism gives rise to dynamic effects that must, of course, be distinguished from purely thermostatic effects and can under no circumstances be interpreted solely in terms of equilibrium thermodynamics. If the rate of attainment of equilibrium of a conformational degree of freedom is very low, it becomes difficult or impossible to reach internal equilibrium in a finite time with respect to this degree of freedom.

2.1 Dynamic Effects of the Conformational Isomerism

Conformational isomers are thermodynamically relevant if the isomer concentrations are variable and the equations of state of the system depend on these variables. Other factors of importance are the boundary conditions and the type of external perturbation to which the system is subjected. The relevance of a conformational degree of freedom must hence always be regarded in relative and never in absolute terms. For instance, the rotational isomers in a small side-chain of a molecule may influence neither the pressure nor the volume of the system. The rotational isomers are in this case mechanically irrelevant. It is, however, possible that the mechanical irrelevance applies only to the case of an adiabatic (isentropic) process and that the same rotational isomers would become mechanically relevant under isothermal conditions. Furthermore, electrical polarization or magnetization may well depend on the rotational isomer concentrations, even if volume and pressure do not. Rotational isomerism is then not mechanically, but dielectrically or magnetically relevant. The response-functions (e.g. dielectric constants, magnetic permeability, mechanical moduli, or heat capacity) for the systems with relevant conformational isomers depend on the ratio of the rate at which the conformational isomers attain internal equilibrium to the rate at which the system is subjected to the external forces (electrical, magnetic, mechanical or thermal). The response-functions assume different values depending on whether the system is stressed rapidly or slowly. This fact is exploited to gain information on the dynamics of conformational isomerism [38–41]. For example, mechanically relevant conformational isomers give always rise to inelastic or viscoelastic effects [37,41].

An internal degree of freedom is referred to as arrested or frozen if the rate at which it attains equilibrium is negligible compared to the rate at which the system is subjected to external perturbations. The internal variable corresponding to a frozen degree of freedom remains constant during the period in which the system is subjected to the external perturbations. Accordingly, the term "frozen" applies primarily to a scale of rates. The rate at which the internal degree of freedom approaches thermodynamic equilibrium depends, however, on the external variables, e.g. on temperature or pressure. As a rule it increases with temperature. If a system is subjected to given rate of perturbation at various temperatures, and if the internal degree of freedom is frozen at lower temperatures with respect to this rate, then there is a gradual thawing of the internal degree of freedom when the temperature is raised. It is completely thawed if the rate at which it reaches equilibrium is so fast that the corresponding internal variable instantaneously keeps pace with the internal equilibrium. Then the degree of freedom is in internal equilibrium during the entire period of external perturbation.

A quantitative description of such situations can be obtained for the simple thermomechanical system whose molecules can assume only two conformational isomers [37]. The thermodynamic states for this system are usually uniquely defined by temperature T, pressure p and the internal variable x, where x is the concentration of one conformational isomer and $1 - x$ is the concentration of the other. The differences between the response functions under conditions of frozen and completely thawed conformational isomerism are given by Davies' relationships [37,42,43],

$$\Delta c_p \, \Delta \varkappa_T = Tv(\Delta\alpha)^2 \,, \tag{2.1 a}$$

$$\Delta c_v \, \Delta k_T = -Tvp^2(\Delta\beta)^2 \tag{2.1 b}$$

where c_p and c_v are the specific heat capacities at constant pressure and constant volume, respectively; $\varkappa_T$ is the isothermal compressibility; k_T is the isothermal bulk modulus; α is the thermal expansivity; β is the thermal pressure coefficient; and v is the specific volume.

Equations (2.1 a) and (2.1 b) are formally identical to the Ehrenfest equations for thermodynamic second-order transformations [4,44]. Hence, they might suggest that the process in which the conformational degree of freedom is thawed is a second-order transition. In fact, however, Ehrenfest's equations and the Davies relationships differ fundamentally. The Ehrenfest equations describe a thermostatic phenomenon. In the Ehrenfest equation that corresponds to Eq. (2.1 a) the differences Δ refer to states $Z[T, p, \hat{x}_I(T, p)$ and $Z[T, p, \hat{x}_{II}(T, p)]$, of two different phases (I) and (II) which are in internal equilibrium*. The Davies relationships, in contrast, describe primarily dynamic phenomena. The differences Δ in Eq. (2.1 a) are valid for a system in an initial state $Z[T, p, \hat{x}_A(T, p)]$, whose response-functions are measured in one case at a very high rate of external perturbation [dT/dt (p = const.) or dp/dt (T = const.) significantly greater than dx/dt, where t is the time] and in the other case at quasistatic perturbation rate [dT/dt (p = const.) or dp/dt (T = const.) significantly less than dx/dt]. In the first case the perturbation rate is so high that $x = \hat{x}_A$ = const. during the period of perturbation, i.e. the conformational isomerism is frozen with respect to the external conditions and in an arrested thermodynamic equilibrium. In the second, quasistatic case, the concentrations of the conformational isomers assume always the equilibrium value $\hat{x}(T, p)$. Under these circumstances the system is always in internal equilibrium.

Incidentally, the thermal expansivity measured at constant temperature and pressure, viz. α_T and α_p, and the thermal pressure coefficients measured at constant temperature and volume, viz. β_T and β_v, remain the same in arrested as well as in internal equilibrium, i.e. $\alpha_T = \alpha_p \equiv \alpha$ and $\beta_T = \beta_v \equiv \beta$. In non-equilibrium, i.e. in the states between the two extremes of arrested and internal equilibrium, one finds in general that $\alpha_T \neq \alpha_p$ and $\beta_T \neq \beta_v$.

---

* In the derivation of Eq. (2.1 b), T and v are considered to be external variables, independent from one another. In this paper the circumflex (^) denotes values for internal variables in internal equilibrium.

The Eqs. (2.1 a) and (2.1 b) apply thus actually to a rate scale; and, in the frequent case of cyclic exposure, to a frequency scale ω. If a thermorheologically simple system is considered [41], the frequency scale can be replaced by a temperature scale 1/T. Steps Δ that satisfy Eqs. (2.1 a) and (2.1 b) appear then in the response-functions for systems of this nature that are measured as a function of temperature at given, fixed perturbation rate. The temperature at which the steps occur depends, however, on the rate of external perturbution. The temperature-dependent thawing of conformational isomers in thermorheologically complicated systems can be similarly observed in the response-functions, but the steps no longer satisfy Eqs. (2.1 a) and (2.1 b). These two equations lose, in addition, their validity with respect to the rate scale if, as is the case of polymers, several mutually independent, internal variables are required in order to uniquely define the conformational isomerism. In this case. Eqs. (2.1 a) and (2.1 b) become inequalities [42,43].

Dynamic relationships such as the Davies equations have no significance whatever if the system is always in internal equilibrium with respect to the conformational isomerism, or if the conformational isomerism is frozen in the temperature range investigated. Since for systems in internal equilibrium the internal variables are functions of the external variables, the former can be eliminated and the functions of state can be represented solely by the external variables. This elimination entails, however, a loss in information, and it is thus frequently advisable to retain the inner variables also in the case of thermodynamic equilibrium (cf. Sect. 2.2). Arrested or frozen internal degrees of freedom have no thermodynamic significance whatsoever and can be ignored.

2.2 Thermostatic Effects of the Conformational Isomerism

The existence of a crystal ↔ condis crystal transition is the main question to be dealt with in this section. This is primarily a question of thermostatics. It is thus assumed that the relevant conformational isomers in the crystal concerned are in internal equilibrium. Conformational isomers that are arrested in the crystalline state can be eliminated from consideration. Conformational isomers that thaw in the temperature range of the transition may substantially distort the character of the transition because it would be experimentally difficult to attain internal equilibrium.

Starting from an ideal crystal the production of conformational isomers is generally equivalent to the creation of defects and holes. In thermodynamic equilibrium one can therefore consider the crystal with conformational isomers analogous to any real crystal above $T > 0$ K as a saturated solution of defects and holes. The number of defects and holes increases normally with temperature. Associated with this change is an increase in volume and, frequently, also in heat capacity. The production of defects and holes with rising temperature leads to one-phase premelting in the sense of Ubbelohde [12]. Nevertheless, Lennard-Jones and Devonshire [13] have demonstrated on a specific example (that does not concern conformational isomerism) that, depending on the ratios of the intermolecular forces, a discontinuous first-order phase transition may set-in after the concentration of defects and holes has exceeded a critical value. The authors tried to relate such phase transition to the melting process. In reality the main feature of melting is lacking in such a process, namely the complete disintegration of the crystal lattice i.e. the disappearance of the long-range order. Instead, in such a transition a crystal phase that is more or less free from defects and holes is transformed into a strongly disordered crystal phase. This is precisely the type of transformation that one would expect in the crystal ↔ condis crystal transition.

The defect crystals of Lennard-Jones and Devonshire possess interstices onto which the motifs of the crystal lattice migrate by leaving a hole. The diffusion, and also the concentration, of the defects and holes are largely independent of one another. In the transition described by Lennard-Jones and Devonshire defect and hole concentrations play thus both a part. Opposed to this, in molecular crystals with defects, predominantly formed by conformational isomerism in the molecules, the defect and hole concentrations are closely coupled. In this case the concentrations of conformational isomers alone suffice to describe the transition.

For simplicity we consider again the thermomechanical simple system whose molecules can form only two conformational isomers (Sect. 2.1). If the two isomers do not differ in energy, the conformational isomerism may still give rise to dynamic effects because in non-equilibrium x can assume any value from $0 \leqq x \leqq 1$. In internal equilibrium, however, equipartition exists between the conformational isomers, i.e. $\hat{x} = 1\text{-}\hat{x} = 0.5$. Consequently, the conformational isomerism has no thermostatic effect unless the isomers differ in energy. An ideal, two-level system exists if the internal energy of the conformational isomers of concentration x is u_0 and that of the isomers of concentration $1 - x$ is $u_0 + \Delta u$ and if u_0 and Δu are independent of pressure and temperature. According to the principles of classical statistical mechanics the specific Helmholtz free energy $f = u - Ts$ (where u is the specific internal energy and s is the specific entropy) is given for this system by:

$$f = u_0 + \Delta u(1 - x) + kT[x \ln x + (1 - x) \ln (1 - x)] \tag{2.2}$$

where k is the Boltzmann constant.

From the condition of internal equilibrium $(\partial f/\partial x)_{T,v} = 0$, it follows that the equilibrium distribution of the conformational isomers is given by:

$$\hat{x} = (e^{-\Delta u/kT} + 1^{-1}), \qquad 1 - \hat{x} = \hat{x}\, e^{-\Delta u/kT}. \tag{2.3}$$

Equipartition of the conformational isomers occurs now only at the limit $T \to \infty$. If Eq. (2.3) is inserted into Eq. (2.2), the fundamental equation of equilibrium thermodynamics for the ideal two-level system is obtained:

$$f = u_0 + kT \ln \hat{x}. \tag{2.4}$$

In internal equilibrium, the internal energy is given by:

$$u = u_0 + \Delta u(1 - \hat{x}), \tag{2.5}$$

the entropy by:

$$s \equiv -\frac{\partial f}{\partial T} = k\left[\frac{\Delta u}{kT}(1 - \hat{x}) - \ln \hat{x}\right], \tag{2.6}$$

and the heat capacity by:

$$c_v \equiv -T \frac{\partial^2 f}{\partial T^2} = k \left(\frac{\Delta u}{kT}\right)^2 \hat{x}(1 - \hat{x})\,. \tag{2.7}$$

The characteristics of the two-level system is that $\hat{x}(T)$ and $u(T)$ tend towards a saturation-value with increasing temperature, and pass through a point of inflection. As a result, the heat capacity initially increases with temperature, but decreases as the temperature increases further (cf. Fig. 2.1). The heat capacity passes through a maximum, the so-called Schottky anomaly, at a temperature of:

$$T_A \equiv \Delta u/k\,\eta\,. \tag{2.8a}$$

The dimensionless factor η is given by:

$$(\eta + 2)\,e^{-\eta} = \eta - 2\,, \quad \text{i.e.} \quad \eta \simeq 2.40 \tag{2.8b}$$

Since $\hat{x}(T)$ is a unique function of temperature, the ideal, two-level system cannot undergo a thermodynamic transformation.

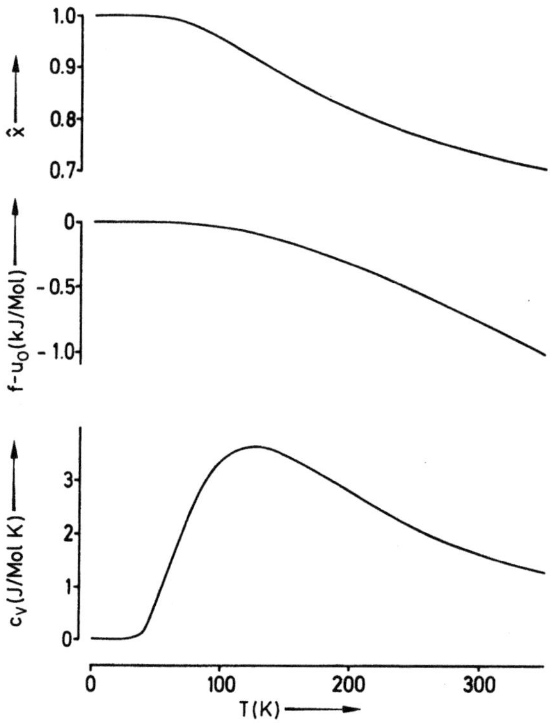

Fig. 2.1. Two-level system in internal equilibrium. Concentration $\hat{x}(T)$ of the low energy conformational isomers, specific Helmholtz free energy $f(T)$, and specific heat capacity $c_v(T)$ as functions of temperature in accordance with Eqs. (2.3)–(2.7). The two-level energy difference Δu = 2.5 kJ/mol

A qualitatively similar behavior is displayed by n-level systems ($n > 2$; finite) in which the energy levels may be degenerated. This category of systems or subsystems includes those with conformational isomers whose energy is governed solely by short-range, intramolecular forces. In principle one does not expect crystal ↔ condis crystal transitions in these systems. Likewise, individual conformational degrees of freedom whose thermodynamic relationships are similar to an n-level system do not make any contribution towards a transition of this nature. Different relationships arise, however, most certainly if, for example, the conformational isomers are subject to steric hindrances that depend on the temperature, pressure or volume; strong dipole-forces or hydrogen-bonds occur instead of weak intramolecular dispersion-forces; or the intermolecular forces play a decisive part. Next, an n-level system will be taken as basis to study in more detail the effect of intermolecular interactions. We shall restrict the discussion to a simple model that represents, at least qualitatively, the relationships in lamellar or bundle-like crystals formed from linear chain molecules such as n-paraffins or polyethylene.

The basic unit of isomerism in linear carbon-chain molecules is the gauche conformation (cf. Ref. [47]). It is obtained by twisting adjacent chain links through a given (relatively fixed) angle. An individual gauche-bond cannot occur in the crystal as a defect for stereometric reasons. Although the formation of gauche-blocks as shown in Fig. 2.2 by cooperative transformation is feasible, it is just as unlikely. The entropy gain in a transformation of this nature is so little that the gauche-blocks can exist only at very high temperatures. The same applies to other stereometrically unfavorable conformational isomers. The only defects that can occur in polymer crystals are those combinations of gauche-bonds that do not appreciably disturb the crystal's bundle or lamellar structure; examples are kinks as described by Pechhold and Blasenbrey [48] or by McMahon, McCullough and Schlegel [49]. These defects are, at least at lower temperatures, statistically distributed over the entire crystal.

The following factors are required to calculate the states of the defect polymer crystal with known, ideal lattice structure:

1. Knowledge of the possible types of defects, their energy, energetic degeneracy, and symmetry; their mobility along the chain, and effective range along the chain;
2. knowledge of the nature and spatial extension of possible defect sites and their possible overlap;
3. knowledge of the interactions between the chain molecules and their dependence on pressure and temperature.

Fig. 2.2. Structure of an n-alkyl lamella in which the chain molecules have individual gauche-bonds. Structures of this nature can only be formed by a cooperative transition at high temperatures. Since the individual gauche-bonds project from the plane of all-trans chains, the structure should actually be regarded as three-dimensional

Fig. 2.3. Various types of contacts between two chain molecules of which both have the same conformational defect, i.e. a kink

The ideal conformational defect gives rise to a perturbation along the entire length of the chain molecule. Although this perturbation may be canceled by other conformational defects, or may attenuate if the chain molecules are very long, it can generally be assumed that a conformational defect remains effective over a long part of the chain [50,51]. The positional correlation of segments in two adjacent chain molecules and the interaction between chain molecules depends thus greatly on the conformation of the chains as a whole. This applies also if the chain molecules have only a single defect of the same type (cf. Fig. 2.3). From a statistical viewpoint it would, therefore, appear necessary to regard the entire chain molecule as a statistical unit and not the individual links or segments. Thermodynamically the defect polymer crystal would then appear to be a multi-component mixture of conformational isomers in which the individual components are interconvertible in quasi-chemical reactions according to the degree of thermal agitation.

To gain an initial, qualitative insight into the relationships that exist in crystals formed from chain molecules, we use a simple model with roughly the same rationale as that of the Lennard-Jones and Devonshire model. Let us regard a bundle of N parallel, stretched chain-molecules as the ideal crystal and let its structure be characterized by a mean coordination number q (Fig. 2.4). This model embraces both, extended chain crystals and lamellae formed by folded-chain molecules. Naturally, in the latter case the "chain molecule" is merely a segment of the real chain. It is assumed that all defects can be built-up from energetically non-degenerate units with

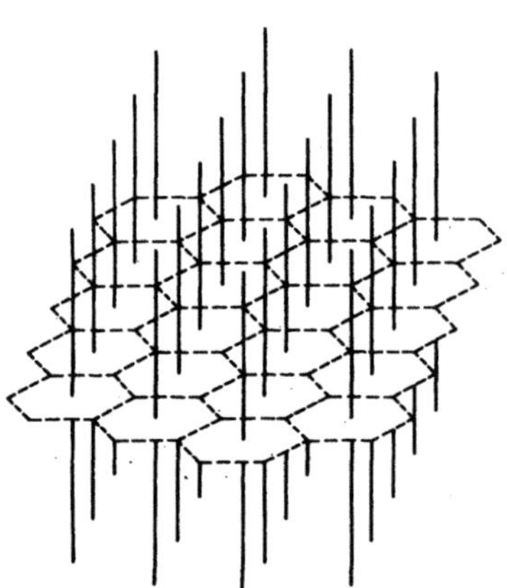

Fig. 2.4. Two-dimensional lattice model for a bundle or a lamella composed of parallel chain-molecules (or chain-segments) with a coordination number of $q = 6$. The angle between the parallel chains and the lattice plane need not necessarily be 90°

an excess energy of Δu. The internal energy u_j of an isolated chain-molecule with j defect units is then given by:

$$u_j = u_0 + j\,\Delta u \tag{2.9}$$

where u_0 is the internal energy of an isolated chain molecule free from defects.

The calculation of the partition function for the ensemble of chain molecules from Eq. (2.9) is closely related to solving the two-dimensional Ising problem with an external field. Since there are no exact solutions for this problem yet, further approximating assumptions must be made to allow simple estimates.

First, it is assumed that N, the number of chain molecules, T, the temperature, and V, the volume or the state of deformation, are fixed (a canonical ensemble). Furthermore, we consider the individual elements of the system, i.e. the chain molecules with j defect units, to be the mutually-independent, statistical elements of the system (this is the Bragg-Williams approximation). In this way the local correlation of the molecules is neglected. The possible types of short-range order are replaced by a single, purely statistical one. The pair-distribution of the chain molecules is then given by:

$$x_{jj} = x_j^2 \quad \text{and} \quad x_{jk} = 2x_j x_k \qquad (j \neq k) \tag{2.10}$$

where x_{jk} is the concentration of the pairs of adjacent chain molecules in which one chain has j and the other k defects.

x_j is the concentration of the chain molecules with j defect units.

The specific Helmholtz free energy f for the ensemble is then given by [51,52]:

$$f = u + kT \sum_j x_j \ln\left(\frac{x_j}{\gamma_j}\right) \tag{2.11}$$

where u is the specific internal energy of the (T, v, ... x_j ...) state,

γ_j is the number of distinguishable conformational isomers in a chain molecule with j defect units.

The valid secondary conditions are:

$$x_0 = 1 - \sum_j{}' x_j\,; \qquad \gamma_0 = 1\,. \tag{2.12}$$

where

$\sum_j$ represents the sum from j = 0 to j = L,

$\sum_j{}'$ represents the sum from j = 1 to j = L, and

L is the maximum number of defect units that a chain molecule can accommodate.

The equilibrium conditions are derived from Eqs. (2.11) and (2.12).

$$\frac{\partial f}{\partial x_j} = \frac{\partial u}{\partial x_j} - kT \ln(\gamma_j x_0/x_j) = 0 \qquad (j \neq 0)\,. \tag{2.13}$$

The equilibrium concentrations for the conformational isomers then follow from Eq (2.13) as:

$$\hat{x}_j = \gamma_j \hat{x}_0 \exp\left(-\frac{\partial u}{\partial x_j}\Big/ kT\right) \qquad (j \neq 0), \tag{2.14a}$$

$$\hat{x}_0 = \frac{1}{1 + \sum_j' \gamma_j \exp\left(-\frac{\partial u}{\partial x_j}\Big/ kT\right)}. \tag{2.14b}$$

Since the energy u depends on all x_j, Eqs. (2.14a) and (2.14b) represent generally a system of L + 1 mutually-coupled, transcendental equations. For $T \to \infty$, the equipartition equivalent is given, as long as the limit of $\partial u/\partial x_j$ remains finite by:

$$\lim_{T\to\infty} \hat{x}_j = \gamma_j \Big/ \sum_j \gamma_j \qquad (j = 0 \ldots L).$$

In order to draw further conclusions from Eqs. (2.14a) and (2.14b), the energy u must be known as a function of the concentrations x_j. The simplest case occurs if the intermolecular interaction can be neglected (ideal system). It then follows from Eq. (2.9) that

$$u = \sum_j u_j x_j = u_0 + \Delta u \sum_j j x_j \tag{2.15}$$

and from Eq. (2.14) that

$$\hat{x}_j = \gamma_j \hat{x}_0 \exp\left(-j\frac{\Delta u}{kT}\right), \qquad (j \neq 0). \tag{2.16}$$

If Eqs. (2.15) and (2.16) are inserted in Eq. (2.11), the equilibrium free energy of the system becomes:

$$f = u_0 + kT \ln \hat{x}_0 \tag{2.17a}$$

where

$$\hat{x}_0 = \frac{1}{1 + \sum_j' \gamma_j \exp\left(-j\frac{\Delta u}{kT}\right)} \tag{2.17 b}$$

Accordingly the qualitative behavior of the ideal system is very similar to that of the two-level system represented by Eqs. (2.4–2.8). If the chain molecule is divided into L nonoverlapping segments, each of which can be occupied by a single defect unit,

$$\gamma_j = \binom{L}{j}. \tag{2.18}$$

Thus Eq. (2.17) becomes:

$$\hat{x}_0 = (e^{-\Delta u/kT} + 1)^{-L} \tag{2.19}$$

and

$$f = u_0 + kTL \ln (e^{-\Delta u/kT} + 1) . \tag{2.20}$$

This is identical to the fundamental Eq. (2.4) for the two-level system if the specific energies f and u_0 are expressed in terms of the mass of the segments that can accommodate a defect unit and not the mass of the whole chain molecule.

An ideal system of linear chain molecules is certainly only fictitious. In reality, the intermolecular forces exert an important influence on the defect concentration and energetics. Since chain molecules with j and k defect units may assume entirely different conformations (cf. Fig. 2.3), the pair concentrations x_{jk} would have to be split up into a sum, viz.

$$x_{jk} = \sum_{l} x_{jk}^{(l)} , \tag{2.21}$$

where the concentrations $x_{jk}^{(l)}$ refers to pairs of chain molecules with the same interaction energy $w_{jk}^{(l)}$. It has been assumed, however, that the chain molecules are statistically distributed. This assumption entails a certain levelling of the intermolecular interactions. Consequently, it appears permissible in the Bragg-Williams approximation to reduce the number of interaction parameters drastically. Only three interaction parameters are allowed for in the following treatment:

w_{00}: for the interaction between two adjacent chain-molecules free from defects;

w_{D0}: for the interaction between a defect chain-molecule and an adjacent chain-molecule free from defects;

w_{DD}: for the interaction between two adjacent defect chain-molecules.

In this case, the only relevant mass variables that occur are the concentration x_0 for the defect-free molecules and Eq. (2.22) for the defect molecules:

$$x_D \equiv \sum_{j}{}' x_j = (1 - x_0) . \tag{2.22}$$

Thus the multi-component mixture is reduced to two components.

The specific internal energy of the system is in this case:

$$u = u_0 + \Delta u \sum_{j} j x_j + \frac{q}{2} (w_{00} x_{00} + w_{D0} x_{D0} + w_{DD} x_{DD}) . \tag{2.23}$$

Nq/2 is the total of possible pairs of chains. The concentrations of the pairs of chains (0; 0), (D; 0) and (D; D) associated with the interaction parameters are given according to Eqs. (2.10), (2.12), and (2.22) by:

$$x_{00} = x_0^2 ; \qquad x_{D0} = 2x_0(1 - x_0) ; \qquad x_{DD} = (1 - x_0)^2 . \tag{2.24}$$

The Gibbs fundamental equation for the system is obtained from Eqs. (2.22), (2.23), (2.14), and (2.11) as:

$$f = u_0 + kT \ln \hat{x}_0 + \frac{q}{2} w_{00} + f_M , \quad (2.25\,a)$$

with the free energy of mixing given as:

$$f_M \equiv \frac{q}{2} \Delta w(1 - \hat{x}_0)^2 , \quad (2.25\,b)$$

and Δw representing the abbreviation:

$$\Delta w \equiv 2w_{D0} - w_{DD} - w_{00} = (w_{D0} - w_{00}) - (w_{DD} - w_{D0}) .$$

If, in addition, Eq. (2.18) applies, the equilibrium concentration $\hat{x}_0$ is governed by the transcendental equation

$$\ln \frac{1-\hat{x}_0}{\hat{x}_0} + \frac{q}{KT} \Delta w \hat{x}_0 = \ln [(e^{-\Delta u/kT} + 1)^L - 1] - \frac{q}{kT} (w_{DD} - w_{00}) . \quad (2.26)$$

It should be remarked that the interaction parameters are normalized such that

$$\lim_{L \to \infty} w_{00} = -\infty .$$

Since conformational defects generally loosen bonds, an energy level diagram of the form shown in Fig. 2.5 is expected. In the determination of the entropy s, the internal energy u and the isochoric heat capacity c_v, it is assumed that the interaction parameters and Δu are independent of temperature.

The equilibrium thermodynamics of the defect crystals is largely governed by the differences in interaction $w_{DD} - w_{D0}$ and $w_{D0} - w_{00}$ or $w_{DD} - w_{00}$ and Δw, where:

$$w_{DD} - w_{D0} = \frac{1}{2} [w_{DD} - w_{00}) - \Delta w]$$

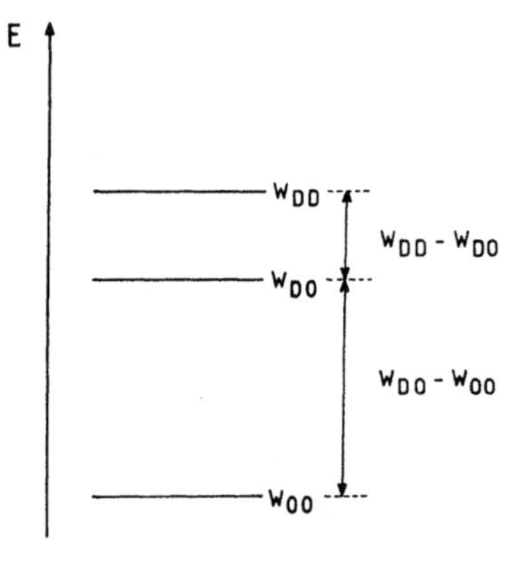

Fig. 2.5. Energy level scheme of pair-bonding between defect (D) and defect-free (0) chain-molecules. Since conformational defects loosen the bonds, it is assumed that w_{DD}, $w_{D0} > w_{00}$. The assumption $w_{DD} > w_{D0}$, as shown, is not a necessary condition

and

$$w_{D0} - w_{00} = \frac{1}{2}[(w_{DD} - w_{00}) + \Delta w] .$$

If w_{D0} is the arithmetic mean of w_{DD} and w_{00}, i.e. $\Delta w = 0$, then w_{DD}-w_{D0} = w_{D0}-w_{00} = $\frac{1}{2}(w_{DD}$-$w_{00})$, and the system of Eqs. (2.25) and (2.26) behaves again qualitatively as a two-level system does. In particular, the heat capacity as a function of temperature passes through a maximum (Schottky anomaly). If, however, w_{DD}-w_{00} > 0, the maximum is shifted towards higher temperatures $T > T_A$ as the difference w_{DD}-w_{00} increases, and the amplitude of the maximum increases (Fig. 2.6). The reason for this behavior is that the intermolecular forces impede initially defect formations but once kT attains a sufficiently high value, the system absorbs as a result of these forces more energy per kelvin than an ideal system.

It can also be supposed that the majority of the conformational isomers assume a mutually favorable stereometric arrangement with a relatively small mean energy difference w_{DD}-w_{00} (cf. Fig. 2.3, left). In this case, the defect chain molecules represent still a considerable perturbation relation to the stretched chain molecules and the following applies:

$$w_{D0} - w_{00} > w_{DD} - w_{D0} \ ; \quad \text{i.e.} \quad \Delta w > 0 .$$

This implies that there is a tendency to aggregation between similar chain molecules. As the temperature rises, "islands" of defect chain molecules form in a "sea" of defect-free chain molecules. The number of (D; 0) pairs in the boundaries is kept as small as possible.

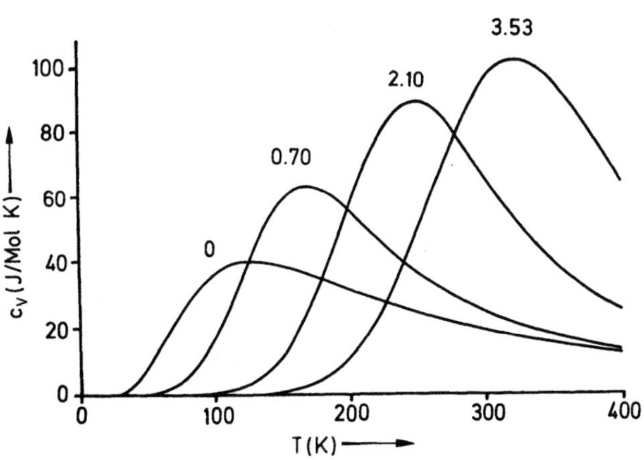

Fig. 2.6. Heat capacity c_v (per mol of chain-molecules) for the cooperative system represented by Eqs. (2.25) and (2.26) in the athermal case of $\Delta w = 0$. The parameter is w_{DD}–w_{00} [kJ/mol]; $q = 6$; $L = 11$; and $\Delta u = 2.55$ kJ/mol. The Schottky maximum for the two-level system (w_{DD}–$w_{00} = 0$) is shifted in the direction of higher temperatures as the bonds between the disturbed pairs of chains are increasingly loosened

If the difference $w_{DD} - w_{00}$ is kept at a constant value and the forces of aggregation $\Delta w > 0$, are made to increase, the Schottky maximum is initially displaced further in the direction of higher temperatures. Its flanks become simultaneously steeper, and the peak height increases. The shape of the temperature/heat-capacity curve for the defect crystals becomes more and more that of a diffuse, but comparatively sharp transformation (Fig. 2.7). The reason for this development is that the initially necessary formation of (D;0)-pairs becomes with increasing possibility of (D;D) aggregation energetically less favorable. Thus, during and after formation of (D;0) pairs, large zones of (D;D) pairs may occur with increasing suddenness.

Ultimately the Schottky maximum becomes severly deformed and when a critical value for the aggregation, Δw_{crit}, is exceeded, the maximum becomes discontinuous on its high-temperature side. The diffuse transformation is then changed into a discontinuous transformation (Fig. 2.8). The reason for this change to a first order transition is the loss of stability of the system [51,53] if

$$q\,\Delta w \hat{x}_{D0}/2 > kT\,. \tag{2.27}$$

The critical value Δw_{crit} and the temperature T_c at which the discontinuous transformation becomes posible are given by the highest, but perhaps only fictitious, value $\hat{x}_{D0} = 1/2$ for the concentration of the (D;0) pairs. According to Eq. (2.24) this

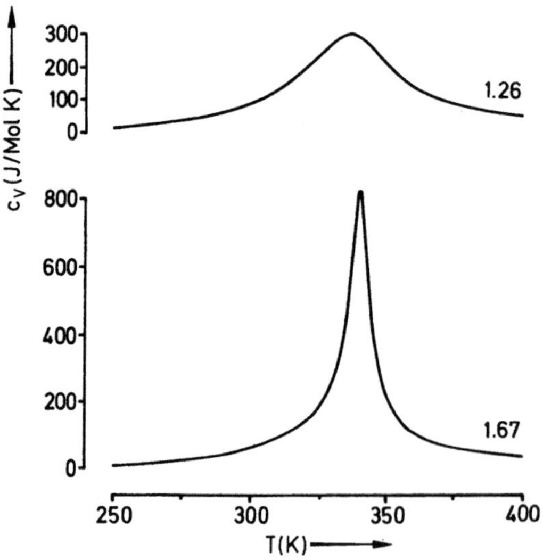

Fig. 2.7. Heat capacity c_v (per mol of chain molecule) for the cooperative system represented by Eqs. (2.25) and (2.26) with the aggregation tendency Δw[kJ/mol] listed as parameter. $w_{DD}-w_{00}$ = const. = 3,53 kJ/mol; Δu = 2.55 kJ/mol; q = 6 and L = 11. As the tendency to aggregation increases, the Schottky anomaly assumes more and more the shape of a comparatively clearly defined, but diffuse transformation

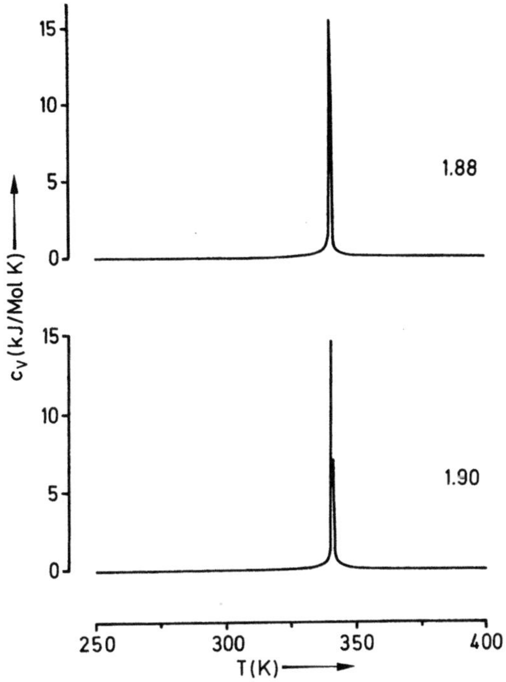

Fig. 2.8. Heat capacity as in Fig. 2.7 showing the approach and exceeding of the critical value for aggregation $\Delta w_{crit} = 1.89$ kJ/mol. The diffuse transformation becomes a discontinuous, first-order transformation

value is reached at $\hat{x}_0 = 1/2$. Inserting $\hat{x}_0 = 1/2$ in Eq. (2.26) yields the following equation for the determination of the transition temperature T_c

$$kT_c \ln [(e^{-\Delta u/kT_c} + 1)^L - 1] = q(w_{DD}\text{-}w_{00})/2 . \tag{2.28}$$

The transition temperature is thus independent of Δw and depends solely on w_{DD}-w_{00}, L, q, and Δu. The discontinuous transformation does, however, only occur at T_c, if additionally the critical value obtained by inserting $\hat{x}_{D0} = 1/2$ into Eq. (2.27) is exceeded, i.e.:

$$\Delta w_{crit} = 4kT_c/q . \tag{2.29}$$

If $\Delta w < \Delta w_{crit}$, the system can still attain the maximum possible value $\hat{x}_{D0} = 1/2$ at or below T_c without losing its stability. If, in contrast, $\Delta w > \Delta w_{crit}$, the equilibrium condition defined by Eq. (26) becomes multi-valued. At T_c, the system reaches only a concentration $\hat{x}_{D0} < 1/2$, at which according to Eq. (2.27) it becomes instable. When T_c is exceeded, the number of defect chain-molecules corresponding to equilibrium can only be produced by an abrupt jump to form extensive (D; D)-clusters with a reduced number of (D; 0)-pairs at their boundaries. The maximum possible value $\hat{x}_{D0} = 1/2$ remains in this case purely fictitious.

With further increase of aggregation above the criticl value Δw_{crit}, the concentration $\hat{x}_{D0}$ of the (D; 0)-pairs reached at T_c decreases continuously. The production of defects in the form of (D; D)-clusters is concentrated more and more on the singular point T_c. As a result, both, the general heat capacity level and its jump

$$\Delta c_v \equiv \lim_{\delta \to 0} [c_v(T_c - \delta) - c_v(T_c + \delta)] \tag{2.30}$$

decreases as the aggregation tendency increases further. If $w_{D0} > w_{DD}$, a step develops in the heat capacity curve, which initially was lambda-shaped in the $w_{D0} < w_{DD}$ range (Fig. 2.9). It needs to be pointed out that for $\Delta w > \Delta w_{crit}$ there is a break in the free energy curve at T_c as defined by Eq. (2.25) and a finite jump in the entropy $[-(\partial f/\partial T)]$. Hence, the transitions are connected with a latent heat. Although the limiting value, as defined by Eq. (2.30) remains always finite, or even disappears (cf. the change from the top to the bottom curves of Fig. 2.9), the heat capacity c_v at T_c is itself infinitely large (not shown in Figs. 2.8 and 2.9). Under the conditions of constant deformation, resp. volume, the crystal ↔ condis crystal transition may appear within the limits of the Bragg-Williams approximation as a normal, first-order transformation. Accordingly, if there are hardly any (D; 0) pairs formed because of high aggregation tendency, and the transformation is practically 100%

$$(0; 0) \xrightarrow[T_c]{} (D; D) ,$$

the heat capacity degenerates to a Dirac delta function:

$$c_v \sim \delta(T_c - T) . \tag{2.31}$$

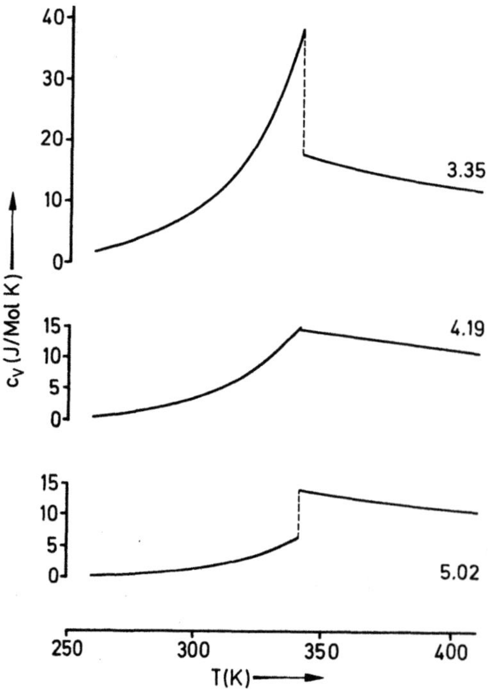

Fig. 2.9. Heat capacity as in Fig. 2.7. At constant difference w_{DD}-w_{00} but increasing tendency of aggregation, the transformation, which is initially lambda-shaped, changes into a step-like transformation

According to Eq. (2.28) the position of the transition temperature T_c depends on the maximum number of chain segments L that can sustain a defect. It can be assumed that L is proportional to the length of the chain-molecules or to the thickness of the lamellae if the chain-molecules are folded. Thus, a relationship exists between the transition temperature and the chain length (or lamellar thickness), i.e. T_c increases with the chain length (lamellar thickness). In general, $T_c > T_A$; in other words, the transition temperature lies above the Schottky anomaly and the discontinuous transformation at T_c develops from the Schottky maximum, as is demonstrated in Figs. 2.6–2.9. If w_{DD}-w_{00} is, however, very small and L is very large, $T_c < T_A$ may also apply, at least in principle. In this case, the discontinuous transformation develops with increasing aggregation on its low-temperature side, independently from the Schottky maximum [51,53].

It must be noted, however, that the energy difference $w_{DD} - w_{00}$ itself depends on the length of the chain molecules. Equation 2.32 certainly applies:

$$w_{00} = Lv_{00} - \Delta v_{00} \tag{2.32}$$

where

v_{00} is the interaction between adjacent segments on two chains free from defects;

Δv_{00} is the correction for the end groups in the chain molecules (if lamellae are formed from folded chain molecules, the end groups consist of loops).

If it is further assumed, as above, that the conformational isomers of two adjacent chains can be arranged stereometrically so that $w_{DD} - w_{00}$ becomes a minimum, it can also be said that, in analogy to Eq. (2.32),

$$w_{DD} = Lv_{DD} - v_{DD} \tag{2.33}$$

where

v_{DD} is the mean interaction per segment between two adjacent defect chain molecules; and

Δv_{DD} allows for the effect exerted by the end groups and, when applicable, also to the effect brought about by a reduction in chain-length.

The relationships expressed by Eqs. (2.32) and (2.33) ensure that the transition temperature T_c does not rise indefinitely, but tends towards a finite, upper limit:

$$\lim_{L \to \infty} T_c(L) = T_{c,\,max} \,.$$

According to Eqs. (2.28), (2.32), and (2.33) this limit is given by:

$$kT_{c,\,max} \ln\left[e^{-\Delta u/kT_{c,\,max}} + 1\right] = q(v_{DD} - v_{00})/2 \,. \tag{2.34}$$

For instance, if $\Delta u = 2.554$ kJ/mol, $q = 6$ and $v_{DD} - v_{00} = 0.422$ kJ/mol, a value of about 400 K will be obtained for $T_{c,\,max}$. Finally, for shorter chain molecules one must also asume [50,51] that:

$$w_{D0} = Lv_{D0} - \Delta v_{D0} \tag{2.35}$$

where

v_{D0} is the mean interaction energy per segment between a defect and a defect-free chain molecule; and

Δv_{D0} is the correction for the effects exerted by the end groups and the reduction in chain length.

It then follows from Eqs. (2.28) and (2.29) that the transition temperature has also a lower finite limit ($T_{c,min}$). Only crystals with chain molecules whose length exceeds a given minimum L_{min} can undergo discontinuous transformation. The value of L_{min} depends on q, Δu and the interaction parameters. The transition temperature increases with chain length, but is subject to the following restriction:

$$0 < T_{c,min}(L_{min}) \leqq T_c(L) \leqq T_{c,max}(L \rightarrow \infty) . \tag{2.36}$$

If Eq. (2.35) is also valid for long chain molecules the tendency to aggregation increases with chain length at constant v_{00}, v_{D0} and v_{DD}. In this case, the heat capacity curve for crystals of long chain molecules is always stepped, as is shown in the lower part of Fig. 2.9; and for $L \rightarrow \infty$, the shape is that of an ideal, first-order transformation, as defined by Eq. (2.31).

The Bragg-Williams approximation with three different interaction parameters states thus that discontinuous crystal ↔ condis-crystal phase transitions are possible in crystals composed of linear chain-molecules if the length of the chain-molecules exceeds a given minimum, and if the defect chain-molecules assume the arrangement of energetically favorable pairs so that a certain tendency to aggregation exists. The transformation is a special case of the Schottky anomaly of the two-level system. The transition temperature increases with length of the chain-molecules, but has an upper, finite limit. Naturally, the Bragg-Williams approximation allows also to compute the energy jump ΔU at the transition point. The value per chain segment $\Delta U/NL$ increases also with chain length and, according to Eq. (2.32) and (2.33) has a finite upper limit [51,52]. It can be concluded from this that the enthalpy of transformation increases also with chain length. The computation of the enthalpy of transformation $\Delta H = \Delta U + V\,\Delta p$ of the canonical ensemble of chain molecules requires, however, a knowledge of the relationship of the excess energy Δu and the interaction parameters with the volume. In experiments, furthermore, the pressure and not the volume is usually given. The states of the defect crystals at constant temperature and pressure can be derived from the specific Gibbs free energy:

$$g = -kT \ln \int e^{-(f + pv)/kT}\, dv$$

and the equilibrium conditions, $(\partial g/\partial x_j)_{T,p} = 0$. Factors that must be known in this case are the relationship between the volume $v = v(T, p, \ldots x_j \ldots)$ and the concentrations of the conformational isomers, and the relationship of the excess energy Δu and the interaction parameters with pressure. An example of how the relationship between pressure and interaction parameters may prove to be a decisive factor is given by polyethylene that can undergo a discontinuous crystal ↔ condis-crystal phase transition only at elevated pressure and contrasts the n-paraffins, whose condis-phase disappears on the application of pressure (cf. Sect. 4.1).

In some cases the Bragg-Williams approximation yields also results that can be evaluated quantitatively, e.g. those concerning transition effects in n-paraffins under atmospheric pressure [51]. In general, however, the information is only to be taken qualitatively. The assumed statistical distribution of the chain molecules imposes certain symmetries on the system. These are reflected, for example, in Eq. (2.24) and by the fact that the transition temperature is fixed by the fictive value $\hat{x}_{D0} = 1/2$. It cannot be expected that these symmetries are of general validity.

In fact, a more complicated pattern of interaction is likely to give rise to different types of short-range order. The symmetries of the Bragg-Williams approximation can be readily broken if, instead of the chain molecules, the pairs of chains corresponding to the interaction $w_{jk}^{(l)}$, described in Eq. (2.21) are regarded as statistical units. This leads to the Bethe-Peierls approximation or to the quasi-chemical approximation of Fowler and Guggenheim [51,54]. The defect polymer crystal appears then to be a substitutional solid solution of conformational isomers that permits, at least in principle, a broad multiplicity of behavior. As in the theory of such solutions, a distinction must then be made between various cases, i.e. ideal, perfect, non-ideal, etc., and between athermal, solvation and aggregation behavior of one or more of the conformational isomers. As in the theory of alloys, superlattices or segregation may occur. In principle, a crystal of this nature is also capable of undergoing several discontinuous phase transformations.

2.3 Summary and Conclusions

Conformational isomerism of molecules may exert a significant effect on the physical properties of a thermodynamic system. This effect depends not only on the structures of the molecules and the molecular ensembles, but also on the nature of the external perturbation and the thermodynamic boundary conditions to which the system is subjected. Consequently, questions about the thermodynamic effect of conformational isomerism should never be expressed in absolute, but always in relative terms.

Since the rate at which concentrations of conformational isomers approach equilibrium is frequently quite low, conformational isomers are often responsible for dynamic effects, e.g. inelasticity or viscoelasticity. Another consequence of low rate is that the internal equilibrium demanded by equilibrium thermodynamics is often difficult to realize experimentally.

In internal equilibrium, conformational isomerism contributes greater heat absorption to the systems. Drastic effects cannot be expected if the sole factors that affect conformational isomerism are short-range, intramolecular forces. If, however, the intermolecular interactions take place, a crystal ↔ condis-crystal transition may appear in the canonical ensemble under certain conditions, either as a diffuse, but very clearly defined transition, or as a discontinuous, first-order transition. Estimates formed within the limits set by the simple Bragg-Williams approximation have shown that a transition of this nature can be expected if there are conformational isomers that tend to cluster. As a general rule, systems consisting of molecules that can form many conformational isomers should be regarded thermodynamically as multi-component mixtures of corresponding complex behavior. Depending on the intramolecular and intermolecular conditions, superlattices or segregations may occur as well as clusters.

3 Condis Crystals of Cyclic Alkanes, Silanes and Related Compounds

Cyclic molecules of small numbers of ring-atoms are rigid and compact so that they can serve as model-compounds for plastic crystals. As the number of ring-atoms increases they develop different conformational isomers. This series of compounds is thus suited to study the interrelation between conformational and orientational dynamic disorder.

3.1 Cycloalkanes

The series of cycloalkanes is the most studied so that the different stages of unfreezing from the crystal state can be identified. The homologous series to be sampled starts with the small, rigid cyclopropane C_3H_6, or perhaps even with ethylene C_2H_4, and continues to very large rings C_xH_{2x} that can have many conformational isomers, and ultimately become a model for chain-folded polyethylene. The section on cycloalkanes is subdivided into homologues of similar phase behavior.

3.1.1 Plastic Crystals (C_2H_4 to C_9H_{18})

The series of homologous cycloalkanes C_xH_{2x} should perhaps already begin with *ethylene* (x = 2). Ethylene is, however, a linear molecule with D_{2h} symmetry. The thermodynamic functions of ethylene are well known from 15 K to the gaseous state with good agreement between statistical thermodynamic calculations for the gaseous state and the experiment [55]. No calorimetric transitions were observed below the melting temperature, 104 K. On fusion, the monoclinic crystals of C_2H_4 (space group $P2_1/n$) gain 32.2 J/(K mol) of entropy [56,57], as is expected for a rigid, non-spherical molecule gaining positional and orientational freedom [7]. At elevated pressure there exists a plastic crystal phase of ethylene, i.e. orientational disorder is introduced below the melting temperature [near the triple point of melt, plastic crystal, and crystal of 110.4 K at 46.8 MPa pressure, ΔS_d is 23.4 J/(K mol) and ΔS_f is 7.0 J/(K mol), as expected for separate gain of orientational and positional dynamic disorder] [58]. The ^1H nuclear magnetic resonance line-width in the plastic crystalline state suggests random molecular reorientation at rates above 10^4 Hz and also a large rate of self-diffusion [58]. In the crystalline state at atmospheric pressure, a 180° flip reorientation about the C=C-axis narrows the NMR-spectrum already above 50 K [59]. The barrier to rotation for this motion was measured from T_1 relaxation time measurements to be 10.9 kJ/mol [59].

The first truly cyclic alkane is *cyclopropane*, a rigid molecule of D_3 symmetry [60,61] with an orthorhombic crystal structure up to the fusion temperature at 145.5 K. Proton NMR detects through second moment and T_1 measurements a sharp decrease of both at about 115 K [62]. Figure 3.1 shows the change in second moment of the ^1H resonance. Good agreement between calculated and measured second moment of the NMR spectrum is achieved with the assumption that at low temperature the crystal is fully rigid. The calculated second moment is then 19.6 G^2, compared to the 20 G^2 measured at 90–110 K [62]. Between 120 and 140 K the measured second moment is about 12 G^2, in accord with the calculated second moment for C_3H_6 averaged about its threefold rotation axis (11.3 G^2). Also shown in Fig. 3.1 is the heat capacity [63].

Table 3.1. Transition temperatures and entropies for cycloalkanes

	T_d (K)	ΔS_d [J/(K mol)]	T_i (K)	ΔS_i [J/(K mol)]
Cyclopropane	no	—	145.5	37.4
Cyclobutane	145	40.8	182.4	6.0
Cyclopentane	122.4	39.8		
	138.1	2.5	179.7	3.4
Cyclohexane	186.1	35.9	279.8	9.4
Cycloheptane	134.8	36.8		
	198.2	1.5		
	212.4	2.1	265.1	7.1
Cyclooctane	166.5	37.9		
	183.8	2.6	288.0	8.4
Cyclodecane	no	—	283.0	72.5
Cyclododecane	199	2.8	333.8	44.4
Cyclotridecane	285	3.2	297.6	25.1
Cyclotetradecane	321.9	51.1	329.3	28.1
Cyclopentadecane	210.1	40.0	336.6	25.2
Cyclohexadecane	268.9	72.8	333.7	22.7
Cycloeicosane	298.0	119.2	323.0	29.2
Cyclotetraeicosane	297.0	127.7	322.0	34.1
Cyclohexatriacontane	no	—	343.0	273.6
Cyclooctatetracontane	no	—	366.0	372.0
Cyclohectacontane	no	—	382.8	633.0
Cyclohexanonacontane	no	—	390.0	821.8

No evidence of any transition is shown below the melting temperature. In addition, the measured entropy of fusion, ΔS_f, is that expected for full orientational and positional disordering [7] (see Table 3.1). It is thus clear that the torsional libration about the C_3 axis must lead to occasional jumps into neighboring symmetry positions that reach the NMR frequency of 10–100 kHz at about 115 K, 30 K below the melting temperature. Such rather large-scale jump-motion does not disturb the crystal symme-

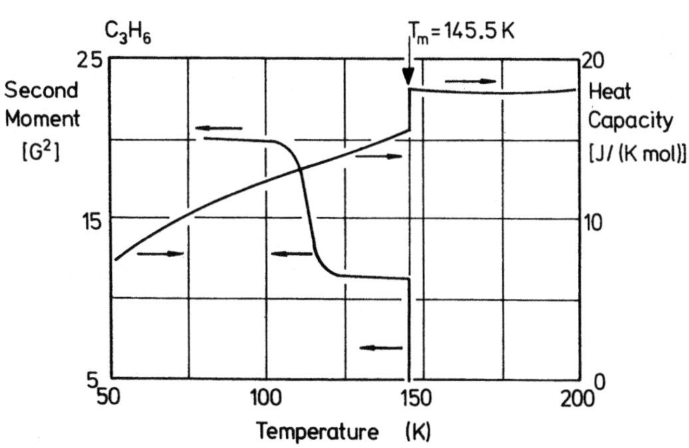

Fig. 3.1. Heat capacity and second moment of the proton NMR spectrum of cyclopropane as a function of temperature. Drawn after data in Refs. [62,63]

try and also does not increase the orientational entropy. In addition, there is no increase in crystal volume beyond what could be taken for normal crystal expansivity. The change in energy of motion is also not detectable within the normal heat capacity precision. At about 100 K the appropriate torsional vibration of about $2 \cdot 10^{12}$ Hz (70–74 cm^{-1}, corresponding to θ-temperatures of 101–106 K) [61] is almost fully excited, i.e. it contributes close to R to the heat capacity and the decrease towards R/2 due to the occasional rotational motion is still small (see Sect. 1.3).

Cyclobutane is now the first member of the cycloalkanes that shows a plastic crystalline state at atmospheric pressure. The molecule deviates by about 15° from planarity, i.e. the molecular symmetry is D_{2d} instead of D_{4h} [64,65]. Above 145 K the crystal structure is body centered cubic (a = 0.606 nm) [66]. In order to achieve such high symmetry, the C_4H_8 must have rotational disorder approaching a spherical average. The distances between the centers of gravity (0.525 nm) are, however, smaller than the v.d. Waals diameter (0.62 nm) i.e. a cooperative, hindered rotation is the only possible motion that leads to such a crystal structure. Table 3.1 lists the thermodynamic transition parameters [67]. Indeed, the entropy of isotropization, ΔS_i, is now decreased to the value expected for positional disordering only, and the disordering entropy, ΔS_d, at 145 K accounts for a typical orientational disordering. Proton NMR absorption spectra and spin-lattice relaxation times have been measured from 115 K to the melting temperature and allow a more detailed interpretation of the motion [68]. Below 130 K the experimental second moment of the absorption line of 20.4 ± 0.6 G^2 suggests effectively rigid molecules (calculated second moment 20.7 ± 1.3 G^2). The likely wagging motion of the out-of-plane CH_2 [69] would have little influence on the second moment. Between 130 K and the transition to the plastic crystal the second moment drops to a narrow temperature range plateau of 11.9 ± 0.6 G^2. Such value (12.3 G^2) would be expected for a reorientation about a fourfold axis. Such reorientation involves, however, a conformational motion of the CH_2-groups on passing through the out-of-plane position dictated by the crystal structure. This is the first example of a "rotoreptation" as discussed in Sect. 1.3. The motion gives rise to a fine-structure in the NMR spectrum just below T_d, due to partial decoupling of the CH_2 of the proton pairs from other proton neighbors, reflected in an additional narrow component of a line-width of 4.7 G. Over the same temperature range the heat capacity shows a pre-transition increase, giving the transition a λ-type appearance [67]. At the transition a discontinuous drop of the second moment occurs to less than 0.12 G^2. Such low value is not explained by isotropic reorientation of the molecules alone (calculated second moment 1.68 G^2), but self-diffusion of the molecules through the lattice at a sufficiently rapid rate must occur to reduce the intermolecular contribution to such small value. From the spin-lattice relaxation times activation energies for the reorientation in the low temperature crystal can be estimated to be 44 kJ/mol and for the self-diffusion, 17 kJ/mol [68]. The self-diffusion seems to occur with little introduction of defects into the plastic crystal, since the disordering entropy in going to the isotropic state is still close to the expected value of 7–14 J/(K mol) (see Table 3.1).

Detailed information on structure and dynamics is also available for *cyclopentane*. Proton NMR studies [70] show in this case coincidence of the linewidth narrowing from a second moment of 26.2 ± 1.5 G^2 expected for a rigid crystal (calculated 28 ± 2 G^2) with a sharp thermodynamic transition [71] at T_{d_1} = 122.4 K. The remaining second moment after the transition is 0.8 G^2 and continues to fall with increasing temperature. The intervening second, small transition at $T_{d_2} = 138.1$ K produces only a minor interruption of this trend (see Table 3.1). Rotation within the molecular plane is insufficient to explain such low second moment. Completely free rotation would eliminate the intramolecular contribution, but still leave 1.4 to 1.7 G^2 from the intermolecular contribution. As for the cyclobutane, a considerable amount of diffusion through the crystal is offered as an explanation for the small second moment that decreases to a value of only 0.03 G^2 at 163 K, about 15 K below T_i). The molecular structure of cyclopentane shows a distinct puckering of the ring. In the mobile phases, and thus also most likely in the plastic crystal phases, this puckering is dynamic, leading to a pseudorotation [34, 72] as discussed in Sect. 1.3. The close-packed hexagonal crystal structure between T_{d_2} and T_i agress with a plastic crystal structure [73], and the low entropy of fusion listed in Table 3.1 may be an indication of the high diffusion rate within the crystal leading to partial positional disorder. The greater puckering relative to cyclobutane seems thus to hinder in-plane reorientation and cause gain of all orientational disorder simultaneously at T_{d_1}.

Nuclear magnetic resonance data on *cyclohexane* [74] are reproduced together with heat capacity information [75] in Fig. 3.2. The transition parameters are listed in Table 3.1. Below 150 K the experimental proton NMR second moment of 26.0 ± 0.5 G^2 corresponds to that calculated for a crystal of rigid molecules of D_{3d} dymmetry in the chair conformation. The decrease in second moment from 155 to 180 K is caused by jump-reorientation about the triad axis with a 46 kJ/mol activation energy. The experimental second moment somewhat below T_d of 6.4 G^2 corresponds to the calculated value of 6.1 ± 1.0 G^2 for such motion. At the transition the second moment drops to 1.4 G^2 which is in line with additional reorientation about all other axes (1.3 to 1.1 G^2 calculated for different assumptions). Above 240 K,

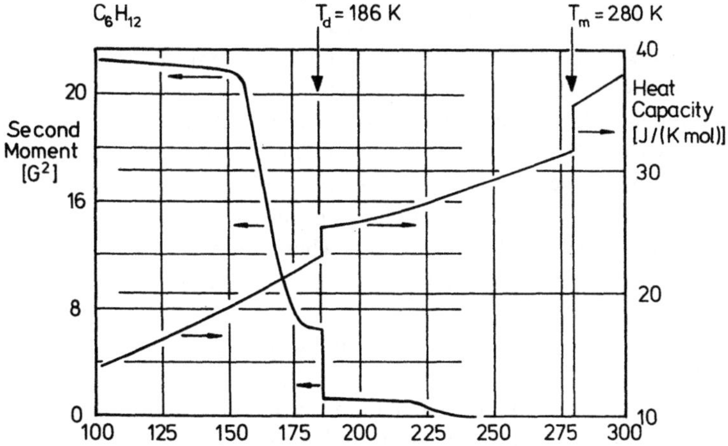

Fig. 3.2. Heat capacity and second moment of the proton NMR spectrum of cyclohexane as a function of temperature. Drawn after data in Refs. [74, 75]

40 K below isotropization, the second moment drops again, now to values of 0.01 G^2. Such low value is, as in the smaller rings, an indication of diffusion within the crystal so that the intermolecular contribution to the NMR line-width goes towards zero. The X-ray structure [76], measured at 115 K (monoclinic crystal, C_c or C2/c) agrees with fixed, rigid C_6H_{12} motifs of close to D_{3d} symmetry. There is only little difference in the structure parameters between 115 and 175 K, excluding any true rotation about the threefold axes. The plastic crystal (form I, cubic, Fm3m) was analyzed at 195 K using the assumptions of isotropic rotation and averaging of a large number of equivalent positions (24). Both give fairly good agreement with the diffraction data [76].

This leaves the question of the possible conformational motion in the plastic crystal phase. Infrared and Raman information [77] suggest that the conformation is largely that of the liquid in which the chair-conformation predominates. For *cycloperfluoro-hexane* it could be shown by multiple pulse (WAHUHA) ^{19}F NMR that in the plastic crystalline phase at 200 K the molecules rotate faster than $6 \cdot 10^4$ s^{-1} while exchanging between the chair conformations with less than 10^3 s^{-1} [78]. At about 300 K, the AB-quartet of absorption lines of the axial and equatorial ^{19}F spins of the chair conformational isomer collapses to the single line of the quickly interchanging species. A comparison with similar data on cycloperfluorohexane in solution shows that the barriers to inversion in solution and in the plastic crystalline state do not differ substantially [78,79]. Furthermore, the solution data reveal that despite larger steric hindrance, the chair-chair inversion rate constants in cycloperfluorohexane and cyclohexane are similar. It seems, thus, reasonable to assume that in cyclohexane as in cycloperfluorohexane solution and plastic crystal may be similar in their dynamic conformational behavior.

The discussions of the conformational isomers of cyclohexane that had their beginning almost 100 years ago with the recognition of the boat-chair isomerism [80], led to a detailed description of pseudorotation in gas-phase and solution (see also Sect. 1.3) [81]. Details were resolved more recently by NMR experimentation [82–84]. The rate constants for the chair to boat-twist conformations in the disordered condensed state are $1.8 \cdot 10^{-7}$, $2.6 \cdot 10^{-2}$, 32, and 3780 s^{-1} for 150, 200, 250, and 300 K, respectively [82]. One expects thus that in the plastic crystalline state the conformational mobility starts without any sharp transition. At temperatures above 240 K there is the additional diffusional motion that increases up to the ultimate, cooperative melting process. The transition entropies in Table 3.1 show no indication of major changes in conformational entropy between the two transitions. Gaining of orientational disorder at T_d and positional disorder at T_i could account for the observed values, although compared to cyclopentane, ΔS_i seems somewhat too large for disordering of a crystal with diffusing motifs and may indicate a discrete increase in conformational motion.

Cycloheptane and *cyclooctane* data on the thermal properties are also given in Table 3.1 [85]. They show little change from the cyclopentane and cyclohexane properties. Again, there is no indication of increasing amounts of conformational entropy in the transition entropies. For cyclooctane in solution ^1H and ^{13}C NMR could prove ring-inversions and pseudorotation among the boat-chair conformations through the twist-boat-chair intermediate to very low temperatures (100 K). Only about 6% of the cyclooctane could be found at about 300 K in the other three crown-family conformers (crown, chair-chair, twist-chair-chair) [86]. These five conformations have the lowest energies of the 10 distinct potential-energy minima [87/88]. Recent solid state ^{13}C NMR (CP-MAS) as a function of temperature showed conformational mobility below T_d [89]. At 110 K the molecules are still rigid, but their conformation does not have the expected high symmetry, it gives rise to six ^{13}C resonance signals. These resonance signals coalesce to one by about 140 K, an indication that within the time scale of the NMR experiment the ring atoms exchange between the different positions of the molecule. The motion below T_d is thus quite different from cyclohexane where rotation started about the C_3 axis without appreciable changes in conformation. Since no disorder is detected by thermal analysis, the molecular motion cannot be a ring-inversion or pseudorotation by itself, but it must be coupled with a rotation to a rotoreptation (see Sect. 1.3).

For all cycloalkanes discussed up to now (with the possible exception of cyclobutane) conformational disorder is only observed in the plastic crystalline state, and even there it develops largely gradually, not much different from the molecules in liquid solution. The number of different, low-energy conformations is rather limited, but high conformational mobility exists through ring inversion and pseudorotation. That conformational mobility in the crystal is possible through rotoreptation was illustrated by cyclobutane and cyclooctane. In Fig. 3.3 a plot of transition entropies per CH_2-group is given as a function of CH_2-groups in the molecule. In this region of plastic crystals, one observes a rapid decrease in total transition entropy (filled circles) as well as the entropy change on disordering transitions (open circles).

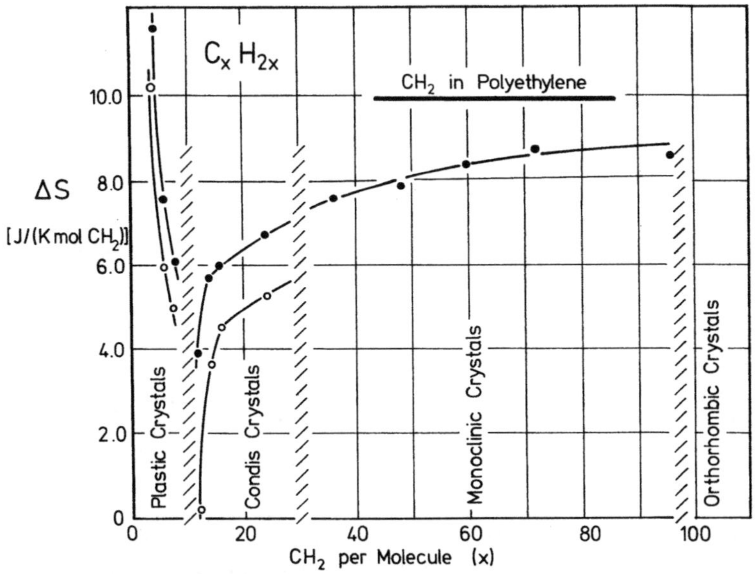

Fig. 3.3. Plot of entropy change per CH_2 of cycloalkanes C_xH_{2x} in transitions as a function of CH_2-groups per molecule. Filled circles: total entropy change (fusion and disordering entropies). Open circles entropy change crystal to plastic crystal ($x < 10$) or condis crystal ($x > 10$). For data see Table 3.1. Also given is the value for polyethylene [7]. To avoid the usual odd-even fluctuations observed in homologous series only even values of x are considered

The reason for this decrease is that both, translational and orientational entropies are size-independent. Both, from Table 3.1 and Fig. 3.3 a change in behavior is obvious for the next group of cycloalkanes ($C_{10}H_{20}$ to $C_{24}H_{48}$). In this next series of cycloalkanes the total transition entropies per CH_2 increase with ring size since the molecules become more disordered. The high temperature crystal phase is now for the larger rings clearly a condis phase with a larger ΔS_i than for plastic crystals. Obviously the cyclic molecules have become too large (and mobile) to rotate as a whole. No plastic crystal phase can be found, and ΔS_f accounts for positional and orientational disordering of the molecule as a whole (see Table 3.1).

3.1.2 Condis Crystals ($C_{10}H_{20}$—$C_{32}H_{64}$)

Cyclodecane ($C_{10}H_{20}$) as the first molecule of this series is exceptional. Just as cyclopropane of the previous Sect. 3.1.1 does not show a plastic crystalline state, there is no disordered phase for cyclodecane. The molecule shows a "normal" melting behavior ($T_i = T_m$). The entropy of fusion is unusually large and suggests positional, orientational and also conformational contributions. In solution the conformational mobility, as determined by NMR, is similarly high as in cyclooctane [90]. In the crystal fast conformational motion exists also, similar to cyclooctane, but it starts because of the lower molecule symmetry at an almost 100 K higher temperature (about 230 K) [89].

The first molecule that shows a disordering transition different from the plastic crystals of Sect. 3.1.1 is *cyclododecane*, ($C_{12}H_{24}$). Only a minor transition is found at 199 K [91]. Infrared [92,93] and X-ray analysis [94] permit the establishment of the ring structure as shown in Fig. 3.4. Also shown in Fig. 3.4 are the ^{13}C NMR spectra for the three condensed phases [91,95]. At 150 K two different carbon atoms can be identified. One bordered by tg and gt bond conformations, representing the

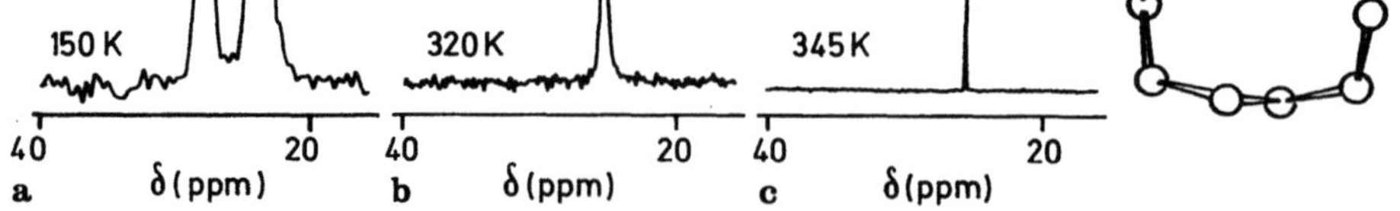

Fig. 3.4a–c. ^{13}C NMR spectra of cyclododecane at different temperatures in the low temperature crystal (**a**), high temperature crystal (**b**), and in the melt (**c**) [91]. The usual line-broadening due to anisotropic shielding is eliminated by magic angle spinning (MAS) and the signal of the natural abundance ^{13}C nuclei is enhanced by cross-polarization (CP). The ring conformation drawn corresponds to the X-ray analysis of the low temperature crystal [94]. The chemical shifts in ppm relative to tetramethylsilane (TMS) are indicated (tg · gt, 4 nuclei = 27.91; gt · gg, = 8 nuclei = 23.58). At 130 K the two spectral lines split further into two and four components, i.e. the molecule becomes less symmetric [89]

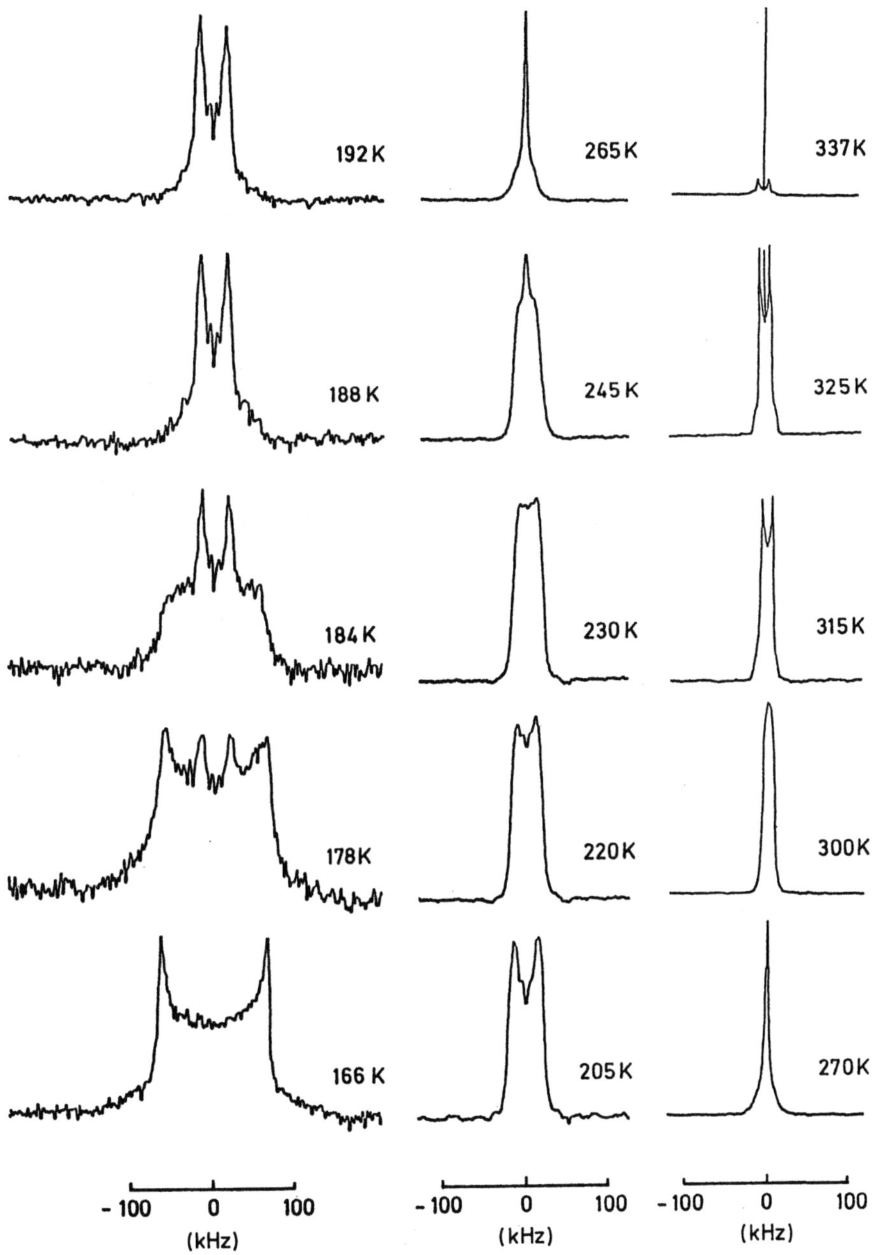

Fig. 3.5. Deuterium NMR spectrum of cyclododecane as a function of temperature as published in Ref. [97]

corners of the molecule, and the other bordered by gt and gg bonds, representing the inner carbons. In this nomenclature t stands for a trans-conformation and g for the gauche-conformation. The four-sided molecule has in the Dale nomenclature [96] that lists the number of bonds within each side of the polygon, the symbol [3333] and is close to a D_4 symmetry. At room temperature the monoclinic crystals (C2/m) randomly is disordered relative to the two enantiomorphs $(g^+tg^+)_4$ and $(g^-tg^-)_4$, in the transition region an incomplete ordering seems to occur with only minor changes in lattice parameters [94]. This accounts for the small ΔS_d of Table 3.1 that is close to Rln 2. The NMR spectrum of Fig. 3.4 requires above the transition temperature a full averaging of the two ^{13}C nuclei within the NMR time scale, not only the exchange between the two enantiomorphs. Thermal analysis and NMR data require thus dynamic disorder. Since within the crystal pseudorotation is not possible, rotoreptation must be the type of motion (see Sect. 1.3). This rotoreptation finds its beginning already below the transition temperature, as is obvious from the deuterium NMR spectra, reproduced in Fig. 3.5 [97]. At 166 K the spectrum has a typical 125 kHz Pake-spectrum, indicative of rigid C—D bonds. Narrowing of the spectrum starts above 170 K, proving the beginning of fast motion in the crystal about 30 K below T_d. No major change occurs at T_d, but between 200 and 320 K a newly narrowed component appears. The second moment of the proton NMR spectrum shows a narrowing from 26 G^2 to 6 G^2 over the temperature range 170–190 K [98] that can also be explained with the beginning of rotoreptation below T_d. Cyclododecane represents thus another example of a crystal with intramolecular motion of a type that leaves the crystal symmetry largely unchanged and whose motion starts without a cooperative, first order transition. In solution similar conformational mobility exists above about 165 K, but because of the missing geometric restrictions of the crystal, pseudorotation is the motion mechanism [90,99].

The thermal analysis of *cyclotridecane* ($C_{13}H_{26}$) seems similar (see Table 3.1) to cyclododecane, but a complex supercooling behavior is observed. Mobility as measured by ^{13}C NMR begins even earlier than in cyclododecane. The changes from the behavior from cyclododecane may be caused by the lower symmetry of cyclotridecane that has a [13333] ring structure [89].

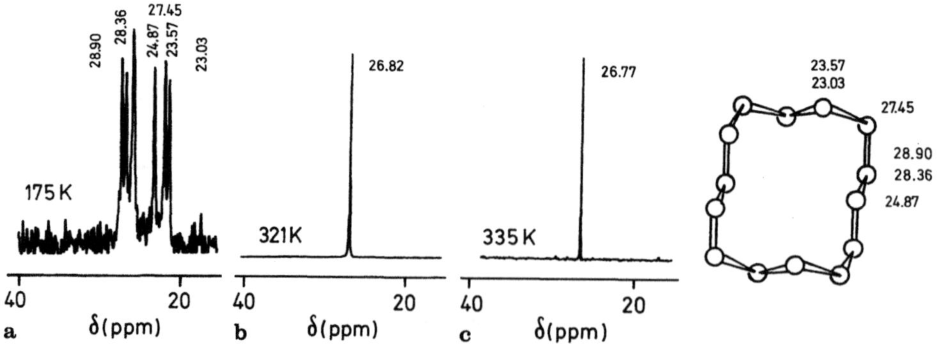

Fig. 3.6a–c. CP-MAS-^{13}C NMR spectra of cyclotetradecane at different temperatures in the low temperature crystal **(a)**, high temperature crystal **(b)**, and melt **(c)** [91]. The ring conformation drawn corresponds to the X-ray analysis of the low-temperature phase [101]. The chemical shifts are indicated in ppm, relative to TMS

With *cyclotetradecane* ($C_{14}H_{28}$) a distinct difference in the size of the considerably elevated disordering transition at 321.9 K is observed. The molecule of structure and symmetry C_{2h} [3434] and it's NMR spectrum are shown in Fig. 3.6 [100]. The crystal structure of the low-temperature form agrees with the conformation drawn in Fig. 3.6 (triclinic, $P\bar{1}$) [101]. At T_d the crystal structure becomes monoclinic with a large decrease in density [100]. The infrared spectra were found to change little at T_d, despite a major change in crystal morphology, suggesting no change in overall conformation [102]. Similarly, Raman spectroscopy confirms practically one conformation for both phases [103]. The high mobility in the crystal phase above T_d is again obvious from Fig. 3.6. In contrast to the cyclododecane there is less mobility below T_d in the solid state, but similar mobility in the liquid. The major narrowing of the deuterium NMR spectrum of cyclotetradecane [97] starts only at about 290 K, while solution proton NMR shows coalescence of the resonances at 153 K [90]. At T_d both ^{13}C and ^{2}H NMR signals narrow discontinuously. If one assumes that 75% of ΔS_d has a conformational origin [36 J/(K mol)], as is often the case in molecular crystals, then one expects as many as 76 distinct positions of the stable conformation of the molecule within the crystal. Conformational jump-reorientations between the many different orientations of the cyclotetradecane ring must account for the mobility and orientational disorder in the high temperature phase. Since the high temperature crystal shows still largely only one conformation, it cannot properly be called a condis crystal. The still rather high orientational order (and low crystal symmetry) does not permit to classify the high temperature polymorph as plastic crystal either. As long as there is no precisely fitting classification for such type of dynamic packing disorder produced by conformational mobility we have included these mesophases in the group of condis crystals.

Cyclopentadecane ($C_{15}H_{30}$) is similar to cyclotetradecane in its thermal behavior (see Table 3.1) and has also a similar ^{13}C NMR spectrum. Motion that exchanges the carbon atoms begins somewhat below T_d (coalescence of the low temperature resonances at about 200 K) [100]. At T_d, however, the infrared spectrum changes significantly. The absorption bands become broader and decrease in number relative to the low temperature crystal [102]. On going to the melt there is, in addition, no further change. Cyclopentadecane has thus clearly dynamic, conformational disorder above T_d, i.e. it is the first true condis crystal of the homologous cycloalkanes.

Cyclohexadecane ($C_{16}H_{32}$) has again a rather high symmetry (D_{2d}) in its lowest energy conformation [4444] [96]. On going through T_d, conformational disorder is introduced, detectable, as in cyclopentadecane, by infrared analysis [102]. In solution, pseudorotation begins again at low temperature. From NMR data [104] it is estimated that at 130 K 90% of the conformation is [4444] which drops to 70% at 300 K, the next best conformation being [3535]. The ^{13}C NMR low temperature resonances coalesce at about 155 K. The pseudorotation of the [4444] ring via the [3544] and [3454] intermediates has an activation energy of 28 kJ/mol and time constants of about 170 and 950 s^{-1} at 142 and 155 K, respectively [104]. Going from crystal to condis crystal, new conformations can be deduced from infrared and Raman data, in contrast to cyclotetradecane [103]. An even larger entropy of disordering ΔS_d is in accord with this observation (see Table 3.1). Similar to the smaller rings there is also in cyclohexadecane an exchange of carbon atoms between the different molecule positions below T_d that can only be explained by rotoreptation without change of symmetry within the crystal. The frequency of this motion is, however, decreasing with increasing ring size. At T_d the line-with of the ^{13}C NMR spectrum decreases abruptly to a value close to that of the melt [89].

A further change in the condis phase behavior is seen in *cyclodoeicosane* ($C_{22}H_{44}$) and *cyclotetraeicosane* ($C_{24}H_{48}$) where now defects are introduced in the low temperature conformation before fast, large amplitude motion is possible [89]. Figure 3.7 shows the molecular structure [105] and ^{13}C NMR data for cyclotetraeicosane. Deuterium NMR shows no fast motion below the phase transition at 297 K for this molecule [97]. The coalescence temperature of the ^{13}C NMR spectrum of cyclotetraeicosane is about 298 K, i.e. equal to T_d (see Table 3.1). Rotoreptation in the low temperature crystals is thus possible, but is considerably slower than in the smaller rings. Figure 3.8 shows the change in mechanism of motion at T_d that permits the much faster ^{13}C exchange in the condis phase. The introduction of defects causes also the shortening of the ring that was observed by X-ray diffraction [106]. Some C^{13} CP-MAS solid state NMR data of lesser resolution [107] make it likely that

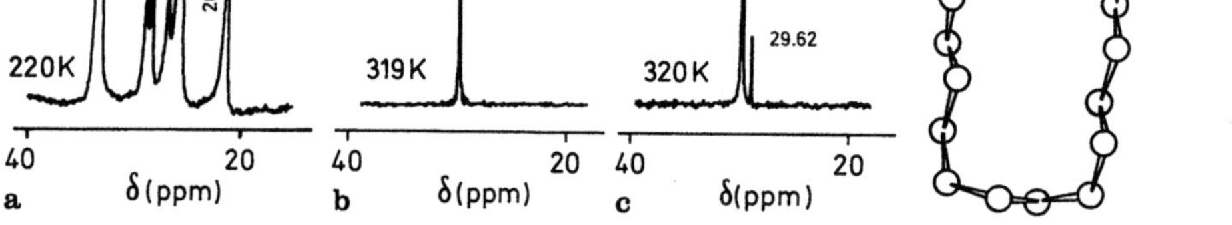

Fig. 3.7a–c. CP-MAS-^{13}C NMR spectra of cyclotetraeicosane at different temperatures in the crystal **(a)**, condis crystal **(b)**, and melt **(c)** [91]. The ring conformation is drawn as suggested in Ref. [105]. The chemical shifts are indicated in ppm, relative to TMS

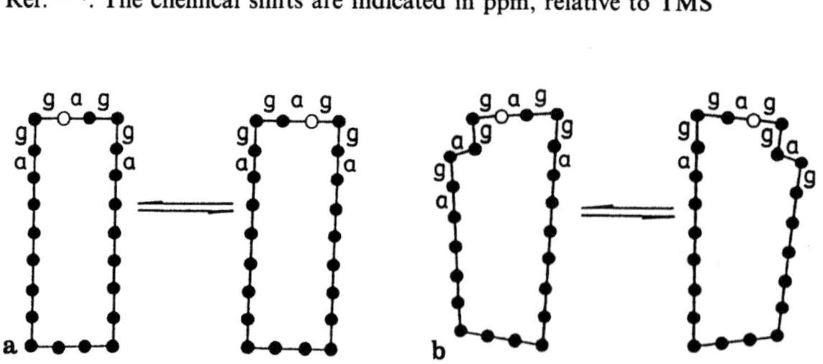

Fig. 3.8a and b. Molecular motion in cyclodoeicosan. **a** In the low temperature crystal by rotoreptation; **b** in the condis phase by defect motion. Drawn after Ref. [89]

similar condis phases are also present in *cyclooctaeicosane* ($C_{28}H_{56}$) and *cyclodotriacontane* ($C_{32}H_{64}$).

3.1.3 Rigid Crystals ($C_{36}H_{72}$ to Polyethylene)

A further change in the behavior of cycloalkanes can be seen with *cyclooctatetracontane* ($C_{48}H_{96}$) where now no phase transitions occur below melting, and motion is detectable by NMR only close to the melting temperature [89,91,97]. Starting with $C_{48}H_{96}$ one observes, however, increasing difficulties in growing perfect crystals.

With even larger rings, such as *cyclodoheptacontane* ($C_{72}H_{144}$) and *cyclohexanonacontane* ($C_{96}H_{192}$), monoclinic crystals that correspond to the crystal structure of the smaller rings [108] can only be grown from dilute solution in n-butyl acetate, for example [91]. On crystallization from the melt, orthorhombic crystals grow. The orthorhombic crystals show as little as half of the ΔS_i listed in Table 3.1 [91,92], but have, depending on thermal history, a more or less broad disordering transition to a hexagonal phase, similar as found in paraffins and polyethylene (see Sect. 4.1) [109]. Depending on crystallization and thermal history, mixtures of both crystal forms may also be produced [89]. The different polymorphs point at a more advantageous fold packing in the monoclinic structure and a more advantageous stem packing in the ortorhombic, polyethylene-like structure [91]. The orthorhombic packing, being closer to the hexagonal arrangement, may also enable an easier change into the condis phase. Only with the cycloalkanes $C_{144}H_{288}$ and $C_{288}H_{576}$ was a close match of the polyethylene lattice achieved that permits no condis phase at atmospheric pressure [110].

Detailed comparison of calorimetry, CP-MAS-^{13}C solid state NMR, X-ray, infrared [89] and also ^{2}H solid state NMR [97] on cycloheptacontane and cyclohexanonacontane resolved the questions of fold packing [89]. In the monoclinic phase the fold is, as in the lower homologues, of a fixed conformation and part of the crystal (ggtgg fold with six identifiable ^{13}C resonances in addition to the tt.tt resonances of the stems, see Fig. 3.7). The fold in the orthorbombic phase, in contrast, is conformationally not well defined, but shows disorder with strain reaching into the trans-sequences. This strain coupled with the orthorhombic packing permits now the development of a hexagonal condis phase [$C_{96}H_{192}$, $T_d = 363$ K, $\Delta S_d = 112$ J/(K mol); $T_i = 388$ K, $\Delta S_i = 449$ J/(K mol) for a sample cooled at 10 K/min from the melt]. Much of the missing entropy (see Table 3.1) is gained already starting at about 200 K through a gradual increase in heat capacity beyond the pure vibrational value.

The motion of the large cycloalkanes can now be understood by comparison of the monoclinic and orthorhombic ^{13}C NMR spectra reproduced in Fig. 3.9. Monoclinic crystals at room temperature are still rigid and of a fold conformation as given in Figs. 3.6 and 3.7 for the smaller rings. The orthorhombic spetrum is considerably broader at room temperature. From ^{1}H NMR spin-lattice relaxation time measurements many disordered, undefined conformations are detected, deviating from the sharp, all-trans signal. At 355 K this broad signal sees motional narrowing and at T_d it becomes abruptly characteristic of the high mobility in the condis state. The monoclinic phase behaves quite differenty. At 355 K the distinct fold conformation is lost and a small orthorhombic signal at 32.8 ppm becomes visible and grows at 375 and 383 K. This means that increasing amounts of mobile, hexagonal

CH_2-sequences are also developing in the monoclinic sample after the low temperature fold-conformation is lost (proof of mobility by ^2H NMR, of symmetry by X-ray diffraction). A monoclinic central part of the all-trans stems with a shift of 34.7 ppm remains in each molecule up to T_m. It initiates the recovery of the monoclinic crystal, on cooling, while after complete melting cooling produces only the orthorhombic phase. The entropy gain of the disordering of the monoclinic phase is completely integrated in the rather broad melting peak.

This detailed study resolves many of the questions of chain folding in polyethylene and its connection to the hexagonal condis phase. In the condis phase all C-atoms of the cycloalkanes can interchange their conformations, i.e. the folds are mobile and permit quick rotoreptation. Similarly, folded chains of polyethylene chain-extend quickly in the condis phase by reptation within the crystal (see Sect. 4.1).

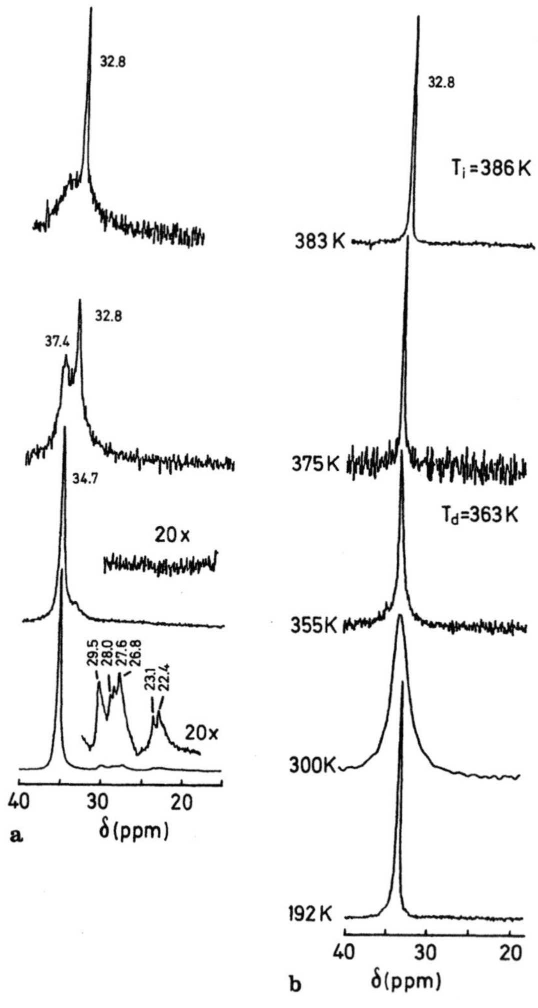

Fig. 3.9a and b. CP-MAS-^{13}C NMR spectra of cyclohexanonacontane at different temperatures for **a** solution crystallized, monoclinic crystals and **b** melt crystallized, orthorhombic crystals. Curves from Ref. [89]

3.2 Substituted Cycloalkanes

A large number of changes, such as addition of side-groups, substitution of hetero-atoms, introduction of double or triple bonds, and bridging of the rings to produce cage molecules, are possible with cycloalkanes. Many of these molecules show dynamic disorder in one or more crystalline polymorphs [32]. A detailed review of over 40 of these dynamically disordered crystalline polymorphs that are mostly plastic crystals can be found in Ref. [46], Chapt. 10. The substituted cycloalkanes permit the study of the effect of lowering the symmetry when compared to the cycloalkanes. Only a few examples have been selected here to illustrate major changes from the cycloalkanes. Little is known about altered large-ring compounds where substitution should hinder rotoreptation (see Sect. 3.1). *Cyclotetradecanone*, $C_{14}H_{26}O$, was analyzed by X-ray diffraction [111] and ^{1}H and ^{13}C NMR [112]. The compound showed at low temperature a regular ring-skeleton structure, but the >CO-group could only be fitted to the X-ray diffraction data by statistically distributing it with different weights over six ring positions. Based on the experience with the cyclo-alkanes, this may point to dynamic disorder at high temperature and may also be a hint that residence times at different positions in the ring are different.

Cyclopentene, C_5H_8, and *cyclohexene*, C_6H_{10}, are examples for cyclic molecules with an identifiable bond. They have been analyzed as to their thermal properties [113] and proton NMR line-width [114,115]. Introducing the double bond into cyclopentane reduces the puckering of the ring to such a degree that the C-atom opposing the double bond is only 0.03 nm out of the plane of the other four carbon atoms. More important, the double bond makes all C-atoms distinguishable. Cyclopentene is, relative to its slightly non-planar structure similar to cyclobutane, but cannot undergo pseudorotation because of the asymmetry. Reorientation about an axis normal to the coplanar carbon atoms begins at about 90 K, 40 K below similar motion in cyclobutane. It is measured by a drop of the second moment of the proton NMR spectrum from 19.6 G^2, that corresponds to the estimated value for the rigid crystal, to 7.6 G^2. The latter second moment would not be in agreement with additional reorientation about axes in the plane of the molecule. Since reorientation within the plane causes disordering due to the double bond relocation, the narrowing of the NMR line-width coincides with an entropy absorption that was not observed for cyclobutane. Before transition to the isotropic state at 138.1 K [ΔS_i = 24.4 J/(K mol)], the NMR line-width narrows at about 124 K to levels explicable only with considerable self-diffusion. This second motional narrowing is not connected with an observable thermal effect, i.e. no additional disorder is introduced. The larger value of ΔS_i shows also that more than just positional disordering must occur at T_i. Note that T_i of cyclopentene is much lower than T_i of cyclopentane, it is even below T_d of cyclopentane (see Table 3.1), a fact related to the higher symmetry of cyclopentane. From the calorimetric results on cyclopentene [113] it seemed that the analyzed sample was below T_d = 87.1 K only partially crystallized since it showed a glass transition at about 54 K. The low ΔS_d of 5.5 J/(K mol) is thus perhaps to be revised upwards. A higher heat capacity of cyclopentene compared to cyclohexane below T_d suggest also less order in the analyzed cyclopentene. The cyclohexene shows above T_d, similar to the cyclopentene, mainly reorientation in the approximate plane of the molecule and no three-dimensional reorientation, as discussed for the corresponding cycloalkanes (see Sect. 3.1, Fig. 3.2, and note the differences in T_d and T_i). As in cyclopentene and in contrast to cyclohexane, the beginning of motion in cyclohexene, signified by the proton NMR line width narrowing, goes parallel with the entropy absorption found by calorimetry [T_d = 138.7 K, ΔS_d = 30.6 J/(K mol); T_i = 169.7 K, ΔS_i = 19.4 J/(K mol), compare to Table 3.1]. As in cyclopentene, it is possible to quench the mesophase to a glass. A second, metastable mesophase and its glass, as well as an isotropic glass, could be produced. The glass transition temperatures were found to be 83, 81, and 78 K, respectively [116]. Conformational disorder is expected in these mesophases which are related to the plastic crystals, but are restricted to reorientation within a plane.

Of the substituted cycloalkanes *cyclohexanol* is perhaps best analyzed [117,118]. Neglecting a crystal polymorph III that is not stable at any temperature, cyclohexanol shows a plastic crystalline polymorph I between T_d = 265.3 K and T_i = 299.0 K with corresponding transition entropies of 33.3 and 6.0 J/(K mol), values typical of a plastic crystal. As with other plastic crystals, polymorph I may be supercooled and forms then a PC glass (see Fig. 1.1) with a glass transition temperature of about 150 K. While cyclohexane showed dynamic reorientation without change in entropy below T_d (starting at about 150 K, see Fig. 3.2), no such motion is possible in cyclohexanol without entropy gain. Dynamic disorder and thermodynamic transition occur together at T_d = 265.3 K when the rigid, ordered polymorph II changes to the plastic crystal polymorph I. The second moment of the proton NMR spectrum drops sharply at T_d from its rigid crystal value of 25 G^2 to about 1 G^2, and somewhat above T_d it drops further, due to the typical self-diffusion of the small cyclic alkanes. In the glassy state (PC glass) the second moment starts decreasing slowly from the same rigid value as found for polymorph II as one approaches the glass transition temperature (150 K). The mobile value of about 1 G^2 is reached already at about 210 K. In the supercooled state the plastic crystalline polymorph may crystallize to the rigid polymorph II with the expected heat of crystallization and decrease in NMR line-width.

An example of globular molecules of similarly different motional behavior is presented by *adamantane* ($C_{10}H_{16}$) and *camphor* ($C_{10}H_{16}O$) [119]. The symmetric adamantane shows motional narrowing of the proton NMR spectrum due to reorientational jump motion between equivalent orientations about its two and three-fold axes starting at 140 K, and shows hardly any further change at the transition to the plastic crystalline state that occurs at 209 K. The asymmetric camphor (d), in contrast, shows the thermodynamic transition to the plastic crystalline state and the motional NMR line-width narrowing together at T_d = 244.5 K. The temperatures of isotropization for the two molecules are 543 and 451.6 K, respectively.

3.3 Other Ring Compounds

Ring-compounds with multiple conformations and conformational mobility do not only exist for organic molecules, as described in Sects. 3.1 and 3.2, but also for many inorganic molecules. There are, for example, many cyclic sulfur-nitrogen compounds that gain multiple conformations as the ring-size increases. Many 4 to 12-membered rings have been made and studied, but little or nothing has been reported about coformational disorder and mobility [120]. Similarly there exists a large number of cyclophosphazenes that could show conformational disorder [121]. The elements sulfur and selenium are known to have cyclic allotropes of sufficient ring size to exhibit multiple conformations [122]. Finally a number of cyclosilanes and cyclosiloxanes have been produced and analyzed [123]. In this section some of these ring-compounds will be briefly discussed to point to the potential source for inorganic condis crystals and some basic differences in conformational energy.

The two major S_8 allotropes of *sulfur* are the orthorhombic α-crystals form (F_{ddd}) and the monoclinic β-crystal form ($P2_1/c$) [123–126]. The orthorhombic crystal is stable only below 368.5 K, but can be heated to fusion at 385 K. The high temperature, monoclinic crystal forms an isotropic melt at 392 K with partial ring decomposition, so that the ideal T_i has been extrapolated to be 406 K [123]. The high-temperature crystal has 1/3 of its molecules randomly arranged in two orientations [125,126]. Adiabatic calorimetry of the supercooled β-crystals seems to indicate that the less dense crystal can gain (or lose) its disorder entropy of about (R/3) ln 2 per mole of S_8 in the temperature range of 120 to 220 K, centered about a lambda-transition at 198 K [127]. The conclusion from these observations must be that the disorder in the high temperature crystal form is dynamic, caused most likely by conformational interchange between the two different orientations of the crown isomers. Another conclusion that seems not to have been checked to date is that below 120 K there must be another, closely related crystal structure of lower symmetry that has only one of the two disordered S_8 orientations remaining. It would also be of interest to find out if in the low-temperature α-crystals and the 2/3 ordered positions of the high temperature β-crystals jump reorientation (or rotoreptation) of the S_8-rings exist. No S^{33} (spin 3/2) NMR of the solid state seems to have been made to resolve these questions. Other cyclic sulfur allotropes with ring-sizes of up to 20 atoms have not been studied in sufficient detail [123], but can potentially give condis crystals in analogy to similar size cycloalkanes.

Of the substituted cyclosilanes *cyclododecamethylhexasilane* $(SiM_2)_6$ has seen most attention. From the physical appearance and mechanical properties of the compound it is quite obvious that the material has a plastic crystal phase of wide temperature-range of existence. Heat capacity and heat of transition measurements fixed the disordering temperature, T_d, at 352 K with a ΔS_d of 47.4 J/(K mol) and the change to the isotropic state at 529 K with a ΔS_i of 7.9 J/(K mol) [11]. Both transition entropies can be justified assuming an ordered molecule becoming a plastic crystal, and then changing to the isotropic melt. Solution proton NMR revealed, however, that the silane ring has high conformational mobility down to the lowest measured temperature (180 K). [11,128]. One would expect thus that such mobility is at least also present in the plastic crystalline state. The crystal structure at room temperature was analyzed by X-ray diffraction [129]. A rather low density, i.e. losely packed, monoclinic (A2/a) crystal structure with motifs on symmetry sites $\bar{1}$ was found. No defects were observed, but a flattened chair-conformation of the molecules showed rather large deviations in the torsional angles from those expected by analogy with cyclohexane. Study of the motion within the crystal was done by proton [130], carbon(13) and silicon(29) [11] NMR. Above about 260 K, i.e. well within the low temperature crystal polymorph, only one resonance line was observed for all three types of atoms. There must thus be rapid interchange between atoms at different locations of the chair conformation. For fixed chair conformations there would have to be two ^{13}C signals, one for the equatorial and one for the axial positions, and, assuming that the $—CH_3$ groups rotate at this high temperature, also two ^{1}H signals. The ^{29}Si should have only one signal because of the symmetry in the chair conformation. An interchange of the atoms of a chair conformation within a crystal is, however, only possible through rotoreptation (see Sects. 1.3 and 3.1.2) since a given chair orientation must be kept in order to avoid disorder. The heat capacity study indicated no transition below T_d and also no detectable gradual entropy gain beyond the vibrational contributions [11]. At temperatures below 200 K multiple-peak spectra are resolved, but instead of the expected two and one absorptions, six and three are found for ^{13}C and ^{39}Si, respectively. Since there is no disordering (or ordering) connected with this change, the fixed low-temperature conformation is either a distorted chair conformation or the "twist" conformation that is well known for cyclohexane, but of higher energy there. In order to keep the crystal structure, the center of inversion $\bar{I}$ must be kept by the molecule. From standard bond-rotation discussions there is no reason for such change in conformation, it must thus be forced by the packing within the crystal that may see a large volume contraction on cooling because of its low density at room temperature (0.91 g/cm^3). The second moment of the proton NMR spectrum remains constant at 6.9 G^2 during the change in the ^{13}C and ^{29}Si spectra. Most of the decrease of the second moment from its calculated rigid value of 29 G^2 is due to the $—CH_3$-rotation that is fast, even at 100 K. Above 200 K this second moment decreases further, to reach about 1.0 G^2 at the transition temperature into the plastic crystal phase. In the plastic crystalline state the second moment changes to 0.3 G^2 and decreases even further at about 300 K when diffusion becomes possible. Again, the diffusion must occur with little gain in disorder. Despite high conformational mobility there is thus little evidence of conformational disorder before reaching the plastic crystalline state. The chair conformation remains the dominant conformation, undergoing jump reorientation and pseudorotation.

Of the cyclosiloxanes, *cyclooctamethyltetrasiloxane* $(SiM_2O)_4$ has seen some attention. Differential scanning calorimetry, ^{13}C and ^{29}Si solid state MAS spectra and some preliminary X-ray data are available [95,131]. A disorder transition T_d occurs at about 265 K with perhaps one third of the total entropy of fusion. The transition to the isotropic state is found at 295 K. In solution the ^{13}C NMR spectrum shows a single line down to the lowest measured temperature (140 K), a sign of fast conformational interchange caused by rather low, rotational energy barriers of the Si—O-bonds. In the solid state, cooling through T_d both, the ^{13}C as well as the ^{29}Si NMR spectrum change to a two-line pattern in accord with the phase transition. The intermolecular forces stop thus the conformational motion and freeze the rings into a tilted crown conformation. At lower temperature the tilt is lost and the molecule becomes more symmetric. It shows then only one ^{13}C and two ^{29}Si NMR resonance lines (crown conformation). There is no indication of $(SiM_2O)_4$ being plastic crystalline in the high temperature phase, it is thus an example of a condis crystal. A similar condis crystal state may also be speculated upon for all larger rings.

Multiple transitions have also been reported for *cyclooctaphenyltetrasiloxane*

$(SiP_2O)_4$ with a possibility of a condis and a plastic crystalline phase (T_{d_1} = 349.8 K, T_{d_2} = 458.8 K, T_i = 478 K) [132,133].

3.4 Summary and Conclusions

In Sect. 3 a set of experimental data on cyclic molecules is described supporting the basic discussions of Sects. 1 and 2. Central are the cycloalkanes that ultimately serve as a model for adjacent reentry, sharply folded polyethylene crystals. Chain-folded polyethylene was shown in the early 1960's to thicken in the crystalline state by straightening as many as 100 to 1000 folds when brought to elevated pressure and temperature. This surprising observation found its explanation in the fast reptation possible in the condis-crystal state.

The following could be documented with the homologous series of cycloalkanes:

1) From ethylene to cyclooctane one finds considerable rotational or rotoreptational jump-motion below the first thermodynamic disordering transition temperature.
2) As plastic crystals with orientational dynamic disorder, the larger rings of this series show also conformational mobility. All plastic crystals develop fast diffusion through a jump-like motion.
3) Cyclobutane is the first cycloalkane that undergoes rotoreptation without major entropy gain, a property that is common for cyclooctane and all higher homologues.
4) From cyclodecane to cyclodotriacontane condis phases replace the plastic crystalline phase.
5) In the transition region from plastic to condis crystalline behavior ($C_{10}H_{20}$ to $C_{14}H_{28}$, see Fig. 3.3) the jump-like rotoreptation leads first to dynamic packing disorder of single conformations by producing, for example, its mirror image. These crystals with packing disorder were, for lack of a finer subdivision, included in the list of condis crystals.
6) Beginning with cyclohexacontane true conformational disorder with considerable entropy gain and a discontinuous increase in mobility at a disorder transition was observed.
7) From cyclohexatriacontane melting occurs in a single step as long as the monoclinic phase is present. Rotoreptational jumps with little or no entropy gain below the melting temperature become slower with increasing ring size.
8) From cyclooctatetracontane orthorhombic crystals with less well defined folds grow from the melt and have a pseudohexagonal condis-phase transition before transition into the isotropic melt. This condis phase is closely related to the paraffin and polyethylene condis phase and has similar dynamics.
9) Only with rings above $C_{144}H_{288}$ seems all rotoreptation to stop and a model of orthorhombic polyethylene been reached.
10) Substitution of the cycloalkanes permits no large amplitude motions without entropy gain. As soon as there is enough interaction energy to order the asymmetric molecules on crystallization, the beginning of large-scale motion is connected with a measurable increase in entropy, usually in a first order transition (Sect. 3.2).

11) Non-carbon, cyclic compounds show similar condis phases, but the intramolecular energy barriers defining the different conformations may be replaced by intermolecular or steric side-group interactions. The prevelant first-order transition marking the introduction of dynamic disorder in most cyclo-carbon compounds change more often to broader, less cooperative transitions (Sect. 3.3).

Overall, the cyclic compounds have formed a rich source of information about condis crystals and plastic crystals and their differences. The limits of condis crystals were reached in crystals showing dynamic packing disorder caused by conformational mobility, but without conformational disorder. The limits of plastic crystals were reached in crystals where the rotational reorienation was restricted within a plane (see also Sect. 5).

With this increased understanding of condis crystals and their limitations, Sect. 4 is designed to illustrate the behavior of flexible, linear macromolecules.

4 Condis Crystals of Flexible Macromolecules

With molecules that are flexible and strictly linear one expects no complication in the phase identification from the liquid or plastic crystalline states. Flexible molecules form random-coil macroconformations on fusion and thus cannot maintain orientational order in the liquid state as is necessary for liquid crystalline phases.* In the solid state the molecules are so long that a rotation of the whole molecule, as required for a plastic crystal, is impossible. It is the goal to describe in this section a larger number of condis crystals of flexible linear molecules. In Sect. 3.3.1 an effort will be made to link the condis phase of polyethylene with the so called "rotor phases" of the n-paraffins. Later sections show the changes in phase-behavior on substitution of the H of polyethylene and introduction of different atom-groupings into the backbone chain. The effects of stiffer chain segments on small and large, linear molecules that may also provide the structure condition for liquid crystals will be analyzed in Sect. 5 and provides the basis for the discussion of stiff, linear macromolecules in Sect. 6.

4.1 Polyethylene and n-Paraffins

Figure 4.1 shows the phase diagram of *polyethylene* [7,134,135]. The well-known, low-temperature orthorhombic crystal structure [136] changes above about 300 MPa pressure, close to the melting temperature into a hexagonal condis phase [137,138]. Figure 4.2 shows the change in lattice parameters accompanying the phase transition. The volume increases by 0.075 cm^3/g (8%) [139,140]. The shortening of the unit cell in the crystallographic c-direction and simultaneous widening at right angles to the

* But note the high temperature phases of polysiloxane (Sect. 4.7). Also α-polyamino acids may form lyophilic liquid crystals after coil-to-helix transition to stiff-chain molecules held by intramolecular hydrogen bonds (see also Sect. 5.3.4).

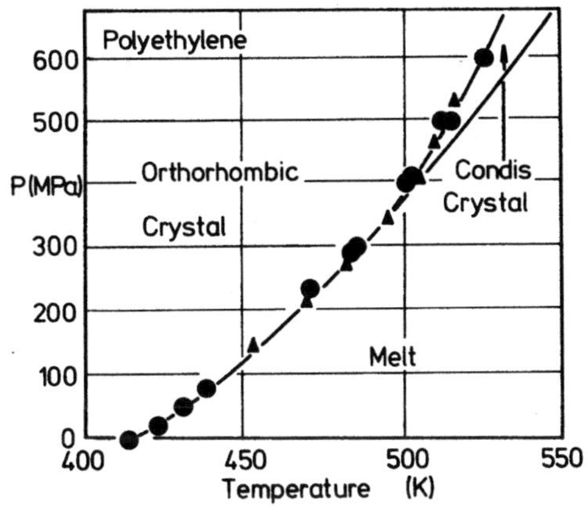

Fig. 4.1. Phase diagram of polyethylene. Typical, broad molecular weight distribution polymer. $\bar{M}_n$ = 10–20,000, $\bar{M}_w$ = 100–200,000. Measured on heating at 0.5–6 K/min. Circles: T. Davidson and B. Wunderlich, J. Polymer Sci., Part. A-2, 7, 337 (1969) and M. Yasuniwa, C. Nakafuku and T. Takemura, Jpn. J. Appl. Phys. 4, 526 (1973). Triangles: D. C. Bassett and B. Turner, Phil. Mag. *29*, 925, 285 (1974). The condis-crystal to crystal phase boundary is drawn according to Refs. [137,140]. For extension to higher pressure see Refs. [134,135]

molecular chain suggests the introduction of gauche-conformations into the backbone chain. Their concentration was measured by Raman spectroscopy using the 1067 and 1135 cm^{-1} skeletal bands. It was close to that found in the molten state [141]. The transition from the orthorhombic to the condis crystal does not involve a reorientation of the molecular chains, since oriented samples keep their c-axis orientation [135]. The condis phase can be stabilized at atmospheric pressure by heating fibrillar crystals under sufficient constraint to reach temperatures above about 423 K (i.e. 8.5 K or more above the equilibrium melting temperature of crystals in equilibrium with relaxed, amorphous melt) [142]. The fibers for such stabilization of the condis phase were grown from dilute solution at 389 K, subjected to Couette-flow and kept at constant length during X-ray and calorimetry experi-

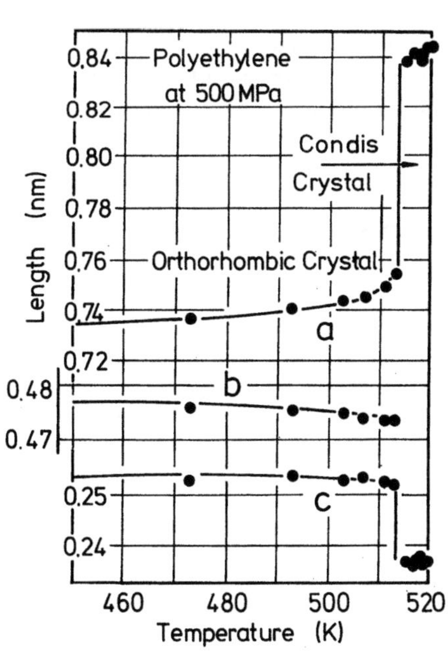

Fig. 4.2. Change of lattice parameters of polyethylene at 500 MPa pressure. Data from Ref. [140]. The condis crystal a-spacing is multiplied with 2 cos 30° to permit a direct comparison with the orthorhombic a-spacing

ments. The enthalpies of transition from the orthorhombic to the hexagonal phase, ΔH_d, were estimated at 500 MPa pressure [140] to be 3.7 kJ/mol, and at atmospheric pressure [142] to be 2.7 kJ/mol. At the higher temperature and pressure condis phase one finds thus a higher entropy of transition ΔS_d [7.1 versus 6.4 J/(K mol), at 520 and 423 K, respectively]. Using for the high pressure condis phase the Clausius-Clapeyron heat of isotropization $\Delta H_i = 880$ J/mol, gained from dilatometry, one can see that 80% of the overall transition occurs at T_d and 20% at T_i. The intermediate ring-size condis crystals of cycloalkanes (see Fig. 3.3) show similar orders of magnitude for the transition heats and entropies.

The especially high mobility in the condis crystalline state was first discussed in the 1960's based on the study of the mechanism of extended-chain crystal growth at elevated pressure from the melt [144–147]. A schematic drawing of the growth-front as deduced from the study of the crystal morphology [147] is shown in Fig. 4.3. To change, for example, a 220,000 MW molecule that crystallizes at the growth-front with about 100 folds per molecule (about 20 nm fold length) to complete extension needs the transport of all 100 folds through segmental motion in the chain direction. Such transport must involve reptation of the whole chain as discussed also for the large cycloalkanes (Sect. 3.1.3). Although linear growth rates were not measured, direct observation by optical microscopy [148] and fast quenching experiments [144] indicate that time scales for complete chain extension of microseconds are not unreasonable. The first suggestion that the hexagonal, high-pressure phase of polyethylene is a mesophase was made by Yasuniva and Takemura [149]. They suggested the phase to be be a nematic liquid crystal. The flexible chain of polyethylene does, however, not contain any rigid mesogenic portions, so that the later classification as a condis crystal is more correct [1]. The X-ray diffraction patterns [135] show on oriented samples sharp 100, 110 and 200 Bragg reflections, indicative of a good two-dimensional lattice, but statistical disorder in the chain-direction as displayed by their meridional reflections. This picture is expected for a condis crystal, not for a nematic liquid crystal [150] which should show only limited one-dimensional order. The dynamic nature of the conformational disorder is proven, besides by the kinetics of chain extension, by ultrasonic studies at elevated pressure [151]. The shear modulus

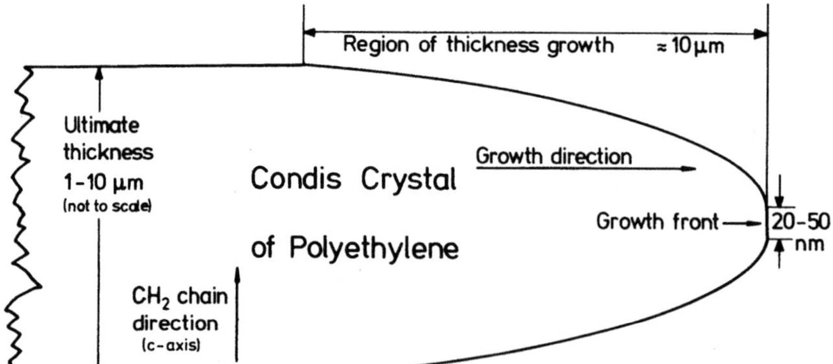

Fig. 4.3. Schematic representation of the growing edge of an extended chain crystal. Drawn after Ref. [147]

of the condis crystal approaches that of the liquid as deduced from transverse sound velocity. In the condis state macroscopic single crystals with low defect concentration can be grown because of this high mobility [152,153]. On cooling into the orthorhombic phase area, these crystals keep their size and become close to perfect [154], (crystallinity above 99%, an otherwise rare occurrence in polymer crystals [47]).

A remaining question [1] involves the reverse reasoning: If in the condis state of polyethylene chain-extension is easy and quick, could it be that during the frequently observed chain-extensions during annealing and drawing or crystallization under mechanical shear conditions the condis state is an intermediate state? It is well known, that on crystallization, metastable states are possible (Ostwald's Rule) and that the catastrophic reorientation on drawing in the neck-region cannot be explained by simple slip along crystal planes [47,155]. Conclusive evidence of the involvement of the condis phase in such processes is, however, not available at present.

Conformational disorder in *n-paraffins* has been proposed for many years. A hexagonal so-called "rotor phase" has been found in a narrow temperature region below ultimate melting [156]. The n-paraffins with even numbers of carbon atoms show the rotor phase for 22 to 44 chain atoms, and with odd numbers of carbon atoms for 9 to 43 chain atoms [157]. It was proposed that this rotor phase is also a condis phase [1]. The importance of this condis phase of the paraffins lies not only in their being a model of the long chain molecules [158], but also in the relevance to the phase behavior in biologically important phospholipids [159].

A series of additional phase transitions of smaller magnitude [157,160] are possible in the paraffins at lower temperatures, dealing mainly with surface and packing adjustments. A detailed description of the rearrangements and motion during these transitions has been given, for example, for tritriacontane, $C_{33}H_{68}$ [161]. These truely solid-solid transitions are not of concern in the present discussion. They are, however, an indication of the importance of the detailed end-to-end packing in paraffins.

The hexagonal condis phase of paraffins has a higher degree of disorder than the orthorhombic, monoclinic, and triclinic crystal phases, but a considerably lower one than the condis phase of polyethylene. From C_9H_{20} to about $C_{21}H_{44}$ the entropy of disordering ΔS_d is about 2.4 J/(K mol CH_2), from $C_{22}H_{46}$ to $C_{36}H_{74}$, about 3.4 J/(K mol) [157]. This is less than half of the ΔS_d found for the condis phase of polyethylene. As a result, the final transition from the condis phase of the paraffins to the melt is much larger than for polyethylene [ΔS_i = 8.1 J/(K mol CH_2) and 7.0 J/(K mol CH_2) for the respective groups of paraffins]. The differences in the heats of transition of the two groups of paraffins are also mirrored in two different types of lamellar packing of the condis crystals [162]. This difference between polyethylene and paraffin condis crystals becomes even more obvious when one looks at the Clausius-Clapeyron expressions for the d and i phase-transitions

$$\left(\frac{dT}{dp}\right)_{\text{transition}} = \frac{\Delta V}{\Delta S} \tag{4.1}$$

For polyethylene $(dT/dp)_i$ is larger than $(dT/dp)_d$, i.e. the condis phase region widens with pressure, as can be seen in Fig. 4.1. For the n-paraffins the picture is reversed, the condis phase region shrinks with pressure, so that above about 300 MPa the paraffin condis phases are not stable any more [163]. Initially this was taken as evidence of basic differences between the paraffin and polyethylene condis phases [164]. It could be shown by Ungar [165], however, that the change from paraffin to polyethylene is continuous. For paraffins with about 40 chain atoms the two slopes are equal i.e.

$$\frac{\Delta S_d}{\Delta S_i} = \frac{\Delta V_d}{\Delta V_i} \tag{4.2}$$

The high pressure condis phase of polyethylene could be observed down to a molecular mass of 6500 [166]. For 1000 and 2000 molecular mass paraffins no condis phases were detected up to 1.5 GPa [166].

For polyethylene, it was possible to stabilize the condis phase at lower pressure, even down to atmospheric pressure, by introduction of defects into the crystal by irradiation or copolymerization with propylene [167]. The transition temperature T_d was found to be a function of radiation dose and lamellar crystal thickness. The defect concentration in the condis crystals at low temperatures was found to be continuously decreasing in the sequence polyethylene, irradiated polyethylenes, copolymers and paraffins with decreasing chain length [167]. Infrared band-intensity measurements led to the following kink concentrations per 100 carbon bonds: $C_{11}H_{24}$, 0.5; $C_{23}H_{48}$—$C_{25}H_{52}$, 1.9; $C_{36}H_{74}$—$C_{40}H_{82}$, 2.3; irradiated polyethylene, 3.2.

The kinked-chain structure is based on a gauche(+)-trans-gauche(—)conformation that leaves the chains parallel and causes a contraction in the chain direction by c/2 with expansions in the a and b directions of about 0.2 nm. The latter may be recovered over several bond-lengths by elastic deformations of the defect region [168].

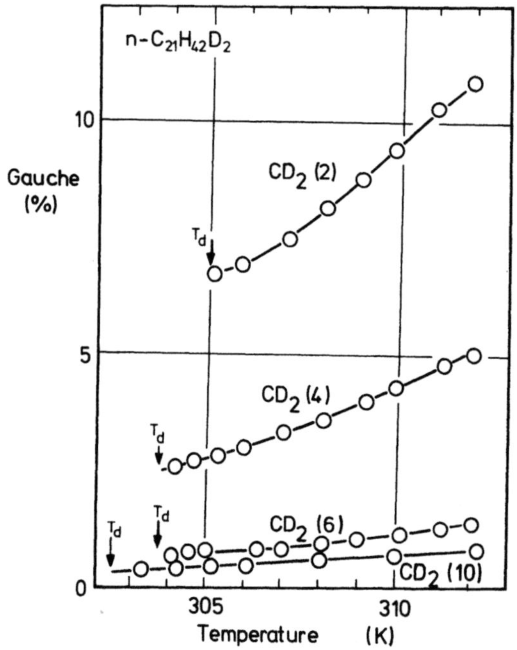

Fig. 4.4. Fraction of C—C bonds in the gauche conformation as a function of position at different temperatures. Data from $C_{21}H_{42}D_2$ deuterated at carbon atom 2, 4, 6 and 10 (curves corresponding to the conformations of bonds 2 or 19; 3, 4, 17 or 18; 5, 6, 15 or 16; and 10 or 11. Drawn after data from Ref. [171])

Electron diffraction [169] and neutron scattering [170] experiments could prove the kinked-chain structure and support also its dynamic nature. Jump rates of about $10^8 \, s^{-1}$ were calculated at 340 K assuming an uncorrelated rotational and translational jump model, ($C_{33}H_{68}$) [170]. The distribution of the defects within the molecules was probed by an infrared CD_2-substitution technique [171]. The concentration of gauche bonds was measured by the intensity of the localized CD_2 rocking vibration frequencies between 670 and 620 cm^{-1} that are sensitive to the conformation of the C—C-bonds adjacent to the CD_2. Figure 4.4 shows the results for $C_{21}H_{44}$ as a function of temperature. The transition temperature into and out of the condis phase are indicated since variously deuterated species show somewhat different transition temperatures. The concentration of gauche-bonds drops exponentially with distance from the ends, but it also increases exponentially with temperature, so that a continuous increase to the high values at the condis phases of polyethylene seems reasonable. Parallel to the increase in gauche-bonds with chain length and temperature goes an increase in lateral unit-cell area, increase in heat capacity, and decrease in activation energy for the gauhe-trans equilibrium. The main cause of the difference in the condis phases of paraffins relative to polyethylene may thus lie in the added effect of the interface.

The difference between the condis states of paraffins and lipid bilayers that have at low temperatures a high gauche concentration (about 20%, half that of an unconstrained lipid) may be found in the difference in lipid and paraffin surfaces (see also Sect. 5.2.3). While paraffins can in the condis state move along the chain-direction to accumulate g and gg defects at the surface, such motion is prohibited in the lipids because of the fixed amphiphilic end of the molecule [171]. A lattice model that seems to properly account for this behavior after introduction of translational disorder for the paraffins [170] is that of Dill and Flory [172].

Proton NMR spectroscopy on n-$C_{19}H_{40}$ [173] and n-$C_{32}H_{66}$ [174] showed the expected low second moment for the condis phase (about 1/4 of the rigid lattice value). Reasonable agreement could be reached with values computed under the assumption of rigid chain rotation. Somewhat more detail could be derived from magic angle spinning ^{13}C NMR studies of the condis phases of n-$C_{19}H_{40}$ and n-$C_{36}H_{74}$ [175].

Table 4.1. Chemical shifts of n-alkanes by MAS^{13}C NMR in ppm relative to TMS

	Central CH_2	—CH_3	α-CH_2	β-CH_2	γ-CH_2	δ-CH_2	ε-CH_2
$C_{19}H_{40}$:							
Rigid[a]	33.4	15.6	25.5	34.9	—	—	—
Condis	33.9	15.3	25.3	35.2	—	—	—
Liquid	31.1	15.1	23.9	33.3	30.7	—	—
$C_{36}H_{74}$:							
Rigid[a]	33.7	15.4	25.7	35.2	—	—	—
Condis	34.2	15.2	24.8	34.5	32.7	33.2	33.8
Liquid	31.2	15.2	24.0	33.4	30.7	—	—

[a] Rigid, but with freely rotating —CH_3 (for the unfreezing of —CH_3 in n-$C_{32}H_{66}$ below 150 K see Ref. [174])

Table 4.1 shows a collection of the observed chemical shifts of the two n-alkanes in the three phases of interest. The 0.5 ppm downfield shifts in going from the rigid crystal to the condis state are caused by changes in magnetic susceptibility on going from the two-chain orthorhombic crystal to the dynamically disordered one-chain hexagonal phase [175]. The reduced downfield shift of the β-CH_2 of n-$C_{19}H_{40}$ and the upfield shift of all other resonances are attributed to the increased gauche conformations. The upfield shifts in CH_2-groups past the β-position in the condis crystals of n-$C_{36}H_{74}$ is in agreement with the infrared data on CD_2-substituted n-alkanes summarized above: longer chain-length and higher temperature produce substantial gauche concentrations further towards the middle of the paraffin lamellae [171]. A rough estimate of the approach of the condis phase to the liquid-like conformation with its approximately 30% gauche conformations led to 65% for the α and β-CH_2 and 44, 33, and 13% for the γ, δ, and ε-CH_2, respectively [175]. A similar estimate based on the CD_2-infrared data results in 50%, 43%, 23%, 16% and 10% for bonds 2 to 6, respectively, in what has to be termed good agreement [171].

The condis state of CH_2-sequences is thus of central importance for the understanding of motion in polyethylene, normal and cyclic paraffins and the related compounds that contain flexible CH_2-sequences in their structure. It will be shown in Sect. 5 that sufficiently long CH_2-sequences attached to mesogens or other rigid structures can similarly give rise to condis phases.

4.2 Polytetrafluoroethylene and n-Perfluoroalkanes

Polytetrafluoroethylene (PTFE) shows a more complicated phase diagram than polyethylene, as can be seen on comparison of Fig. 4.1 and 4.5. PTFE has four crystalline polymorphs in the pT-range of interest [7,176]. Polymorph I has been identified as a condis phase [177]. It has special calorimetric [178] and rheological

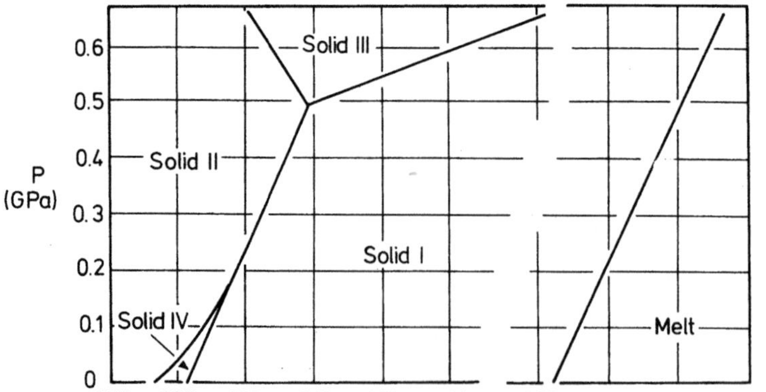

Fig. 4.5. Phase diagram of polytetrafluoroethylene. Diagram drawn after data reviewed in Ref. [7] with an update for the pressure dependence of T_i as given in Ref. [176]. Solid I represents the condis phase

properties [179], and its structure data by X-ray diffraction [180] as well as energy calculations [181,182] and studies of molecular motion by ^{19}F and ^{13}C NMR [183-185] are in accord with such assignment.

Calorimetrically one finds two overlapping endothermic transitions at 292 and 303 K with transition entropies of 2.45 and 0.44 J/(K mol) [177,178]. This compares with a transition entropy to the isotropic melt at 605 K of 6.8 J/(K mol) [177]. The low-temperature polymorph II is triclinic (P1) and consists of domains of either right-handed or left-handed 1*13/6 helices. At 292 K the helices open-up to a 1*15/7 conformation that can be thought of as being generated by 3_1 or 3_2 screw axes operating on the appropriate $(CF_2\text{-})_5$ helical segments [180]. With the added motif and crystal symmetry match, the space group changes from P1 to $P3_1$ or $P3_2$ with a volume increase of 1.3% [186]. In the following disordering transition at 303 K this new helix changes with almost no further volume change (<0.1%) to the condis crystal (polymorph I) [186].

The condis crystal shows only equatorial X-ray diffraction spots, indicative of a regular (hexagonal) packing between the chains, but without matching of the CF_2-motifs along the chain to any regular helix. Calculation of the energetics of the isolated polymer chain yields four rotational isomers per C—C bond [187]. Because of F—F steric hindrance along the chain, the strictly all-trans conformation is hindered by about 1.7 kJ/mol relative to two symmetrical trans-minima rotated ±15° about the C—C bond [t(+) and t(—)]. The gauche-conformations g(+) and g(—) are located at ±123° and have a potential energy higher by 4 to 5 kJ/mol than the trans-conformations. Appreciable concentrations of g(+) and g(—) are, because of the much higher potential energy than in polyethylene (1.6 to 3.0 kJ/mol), expected only at much higher temperature. Packing of the helices built-up of t(+) and t(—) conformers could be shown to be almost independent of the rotation angle between helices [182].

On the basis of NMR and X-ray data [180] it was proposed that torsional oscillations and hindered rotations about the backbone of the molecule lead to the untwisting followed by the disordering of the helix. From the changes in line-shape of the ^{19}F chemical-shift NMR spectrum it was possible to calculate that the rotational diffusion about the chain axis has an activation energy of about 200 kJ and is of similar nature over the complete transition range [185]. At 287 K this motion has a time constant of 50 Hz, at 301 K this has increased to 2000 Hz [185]. Helix reversals [t(+)-sequences followed by t(—)-sequences along the chain and vice versa] are assumed to be collected in the polymorphs II and IV in well defined domain boundaries [180]. Their movement is proposed to explain the phase II to IV transition [186]. Direct evidence for helix reversals was found in the 601 cm^{-1} Raman band that appears during the II to IV phase transition [188]. At increasing temperatures the number of helix reversals increases and leads at about 400 K to equal t(+) and t(—) populations throughout the crystal [189].

The rheological properties of polytetrafluoroethylene had led to the initial identification of polymorph I as a mesophase [179]. Its highly anisotropic flow occurs along planes parallel to the molecular axes and follows the behavior of low molecular mass, smectic liquid crystals. Plots of the logarithm of the shear stress versus the logarithm of the shear rate show a slope of 1/4 that increases only in the melting region. Newtonian flow would show a slope of 1.0, elastic deformation one of zero.

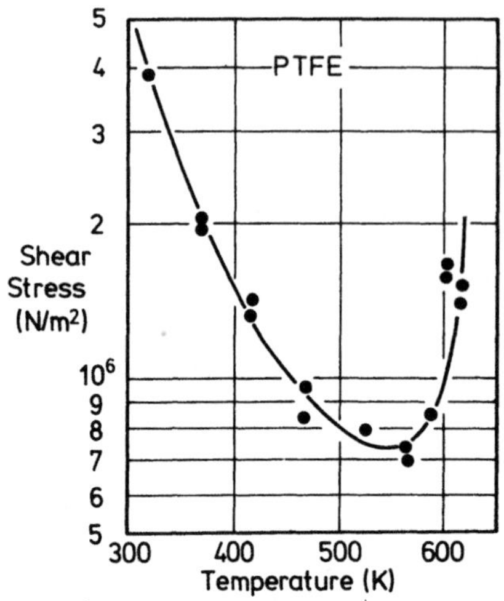

Fig. 4.6. Dependence of the shear stress of condis crystals of polytetrafluoroethylene (crystal form I) as a function of temperature. Drawn after Ref. [179]. Apparent shear rate 8.9 s^{-1} (capillary rheometer, diameter 1.111 mm, length 19.0 mm)

Figure 4.6 shows that below T_d and above T_i the apparent shear viscosity expressed by shear-stress/shear-rate takes on very high values. Below T_d the polymer could not be extruded within the limits of the apparatus and above T_i PTFE of sufficient molecular mass is known to have melt viscosities of above 10^{10} Pa s [179]. Below T_d polymorph II is a rigid crystal with practically only vibrational motion [177] and above T_i the melt viscosity is high because of entanglement of the relatively stiff long-chain molecules.

High mobility of the condis phase was also deduced from the observation of chain extension of PTFE immediately after chain-folded crystal growth or on later annealing [190,191]. The morphology of melt-crystallized PTFE is similar to that of condis crystals of polyethylene, their growth-mechanism is similar to that shown schematically in Fig. 4.3.

Table 4.2. Transition temperatures and entropies for n-perfluoroalkanes[b]

	T_d (K)	ΔS_d [J/(K mol)]	T_i (K)	ΔS_i [J/(K mol)]
CF_4	76.2	22.6	89.6	7.9
C_2F_6	103.9	37.3	176.6	15.8
C_3F_8	99.3	35.6	125.4	3.8
C_4F_{10}	129	0.7	144	53.2
C_6F_{14}	103	10.0	185	36.8
C_7F_{16}	180.5	36.9	221.8	31.3
C_8F_{18}[a]	190/200	35.9	250	41.6
$C_{12}F_{26}$	?	?	347.6	69.0
$C_{16}F_{34}$	103	?	?	?
$C_{20}F_{42}$	190[c]	?	435.6	105.6
$(CF_2-)_x$[a]	292/303	2.9	605	6.8

[a] Sum of two transitions;
[b] Data for C_1–C_8 as collected in Ref. [192];
C_{12} and C_{20} in Ref. [196]; C_{16} in Ref. [195]; PTFE in Ref. [177];
[c] Estimate given in Ref. [109]

The low molecular mass *n-perfluoroalkanes* show also all at least two crystalline polymorphs [192]. The low temperature crystals are expected to be rigid crystals, while the high temperature polymorphs are mesophases. Table 4.2 and Fig. 4.7 show that there are clearly four types of mesophases.

In the first group of n-perfluoroalkanes the mesophases are plastic crystals (CF_4 to C_3F_8). The entropies of disordering, ΔS_d, and of transition to the isotropic melt, ΔS_i, decrease steeply with number of CF_2-groups since both are largely size-independent. The magnitude of ΔS_d corresponds to the entropy of orientational disordering and ΔS_i, of positional disordering. Overall, fusion is thus well separated into two steps.

As the molecules become too long for complete orientational freedom (C_4F_{10} to C_6F_{14}), there is a sharp break in the curves of Fig. 4.7. For this second group of n-perfluoroalkanes ΔS_i includes most or part of orientational disordering. The mesophase may in this case show only motion of the whole molecule rotating about its long axis [192]. The even for such rotation too low ΔS_d for C_4F_{10} could be explained by a largely jump-like motion that does not show significant disorder. Perfluoroalkanes up to such lengths have even in the gas phase, at room temperature, still largely an almost planar zig-zag carbon-backbone, as was documented by electron diffraction [193].

The third group of n-perfluoroalkanes (C_7F_{16} to C_8F_{18}) shows again a large

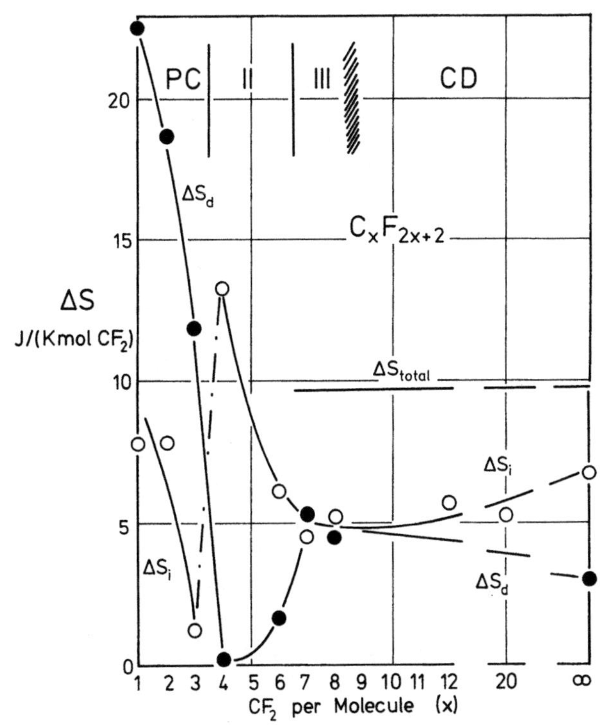

Fig. 4.7. Plot of entropy change per CF_2 of n-perfluoroalkanes in transitions as a function of CF_2-groups per molecule. For data see Table 4.2

disordering entropy and a relatively high T_d (see Table 4.2). In this case, the beginning of conformational disorder is expected because of complete lack of a discrete infrared lattice-spectrum [192]. Compared to polytetrafluoroethylene, the transition entropies in Fig. 4.7 are still relatively large, so that one may speculate that there are still contributions from orientational disorder resulting from rotation of the whole molecule about its long axis.

The somewhat sparse data on the longer chain-length n-perfluoroalkanes in Table 4.2 and Fig. 4.7 [194–196] let one conclude that there is a continuous approach to condis crystals of the nature of PTFE above C_8F_{18}. The infrared spectra of $C_{14}F_{30}$ with a much lower T_d than C_8F_{18} give indications of an ordered helical structure below T_d and random degrees of lateral slip above [194]. The low T_d in this group of compounds is documented in Table 4.2 for $C_{16}F_{34}$ and X-ray diffraction at room temperature was clearly indicative of a PTFE-like condis phase [195]. The entropies of transition to the isotropic melt, finally, of $C_{12}F_{26}$ and $C_{20}F_{42}$ extrapolate smoothly to the ultimate PTFE limit [196].

4.3 Other Fluoropolymers of Type CH_2-CXY

For three macromolecules of type CF_2-CXY- information on conformational disorder and mobility is available. These three polymers are *poly(vinylidene fluoride)* (PVF2), *polytrifluoroethylene* (P3FE) and *polychlorotrifluoroethylene* (PC3FE). Conformational disorder is helped in these three polymers because of a relatively large amount of structural defects. In addition to the regular head-to-tail sequences along the chain, typically 10 to 15% head-to-head defects exist in such samples. The molecules P3FE and PC3FE are, in addition, usually atactic. The reason that such irregular chain molecules can crystallize at all is connected to the isomorphism or isodimorphism of all fluorinated polyethylenes [7 47]. The van der Waals radii of —H and —F in PVF2 and P3FE and of —F and —Cl in PC3FE are not sufficiently different to prohibit cocrystallization. They are different enough, however, to enhance conformational disorder.

In PVF2 the major condis polymorph is orthorhombic (Cm2m) and metastable at atmospheric pressure (polymorph I, also called β). The polymorph results on drawing of melt-crystallized PVF2 or on crystallization at elevated pressure. Its chain-conformation is close to planar zig-zag with randomly disordered rotation-angles of about $\pm 7°$ [197] and occasional kink-defects [198]. There is no direct transition to the more stable polymorphs II (also called α) or III (also called γ). For the transition to these more stable polymorphs, trans-to-gauche changes are needed in large numbers which seems only possible by melting, followed by recrystallization (see also Sect. 4.4) [7]. The condis polymorph I is, however, stable and conformationally mobile at elevated pressure and temperature so that at 350 MPa extended chain crystals can be grown [199]. Because of the need to chain-fold on initial crystallization [47], chain-extension must occur as a secondary process in the condis phase, as outlined for polyethylene (see Sect. 4.1). This is an indication of large-scale conformational mobility under such conditions. It was suggested that the mobile condis phase freezes on lowering of the temperature (and reduction of pressure) so that at room temperature polymorph I of PVF2 is in the glassy state (CD glass, see Fig. 1.1) [1].

For P3FE the condis phase is the only one known. The crystal structure is hexagonal and closely similar to PC3FE. The polymer was considered to consist of helices with sufficient irregularities to have a "smectic-type" structure [200]. The conformational disorder in the crystal was shown to be an irregular succession of tg^+, tg^- and tt-sequences [201]. Again, at room temperature there is no indication of conformational mobility of this polymer. In fact, all of the stable and metastable crystal polymorphs of the fluoropolymers discussed in this section show, up to about room temperature, heat capacities that can be calculated solely from their vibrational frequency spectrum. This is proof that there is no large-scale, disorder-based mobility in these macromolecules [202]. The especially broad glass transition range of semicrystalline P3FE begins at 230 K and ends at about 400 K at a heat capacity that seems to be identical for crystal and melt. It may thus be an overlapping glass transition for both, the amorphous as well as the condis glass, with 400 K being the lower temperature limit of existence of the condis crystal [203]. The transition to the isotropic melt at 495.2 K has a distinctly lower entropy than expected for a fully ordered crystal [ΔS_i – 10.98 J/(K mol)]. It is thus surprising that there has been no report on extended-chain crystal growth of P3FE.

Turning to PC3FE, one finds again extended chain crystal growth [204]. At as little as 10 MPa pressure, extended chain crystals with close to 100% crystallinity have been observed on crystallization from the melt. This observation implies sufficient conformational mobility to permit major rearrangement in the condis state. The crystal structure is, again, hexagonal with a shallow 2*17/1 helix [204]. The transition to the isotropic melt at 493 K has almost the same entropy of transition [ΔS_i = 10.18 J/(K mol)] as P3FE, about half of what is expected for complete conformational disordering (see Sect. 1.1).

While for P3FE and PC3FE only one polymorph has been reported, PVF2 shows the three different major polymorphs introduced above. Besides the condis polymorph I (β), the common polymorph II (α) results on crystallization from the melt below about 425 K [orthorhombic ($P2_1c$) with tg(+)tg(—) 4*1/1 helices] [205], and polymorph (III) (γ) can be obtained from polymorph II on high-temperature (and pressure), solid-state annealing [monoclinic (C_c) with $t_3g(+)t_3g(—)$ 8*1/1 helices] [206]. No direct, fast reversible interconversion between these crystal-forms has been observed. Although double melting peaks for mixed polymorph samples have been recorded, it is not fully resolved if polymorph III is a high-temperature stable crystal, or the overall stable polymorph, so that polymorph II is only kinetically preferred on melt crystallization [7]. All reported entropies of fusion are low [T_i = 483.2 K, ΔS_i = 6.7 J/(K mol)] [203] and indicate considerable conformational disorder. The sluggishness of interconversion between crystal polymorphs indicates, furthermore, that the conformations are largely frozen, i.e. the polymorphs are CD glasses (see Fig. 1.1).

Calculations of the conformational energy minimum of the isolated PVF2 chain by molecular mechanics indicated broad rotation-angle ranges with little change in potential energy. Close to the tgtg(—) helix conformation, rotation angles between 40 and 80° are permitted for g within ±1 kJ/mol [207]. The actual minimum energy is thus largely influenced by crystal packing.

Thermal analysis of the semicrystalline polymorph II of PVF2 led to the proposal of three levels of glass transition-like unfreezing of conformational disorder [203].

First, there is an indication of a condis glass transition in the heat capacity at 360 K that coincides with a dielectric α_c-transition. The α_c-transition was linked to a conformational mobility that converts a PVF2 with fluorine atoms pointing "up" to a chain with fluorine atoms pointing "down" and vice "versa" [208]. Naturally, the whole chain could only be turned upside-down on fusion, followed by recrystallization. Conformational mobility accomplishes this change locally by rotating a tgtg(—)-sequence as shown in Fig. 4.8A. Such change leads to conformational disorder junctions on both sides of the shifted sequence and orientational disorder between the chains. The projection on the ab-plane is not changed during this motion. X-ray data show gradual changes in the crystallographic c-direction with temperature as would be connected to a temperature dependent, dynamic "up" and "down" equilibrium above about 370 K [209, 210].

A second level of conformational disorder concerns the interchange of PVF2-chains with CF_2-groups pointing predominantly to the "left and up" to chains with CF_2-groups printing predominantly to the "right and down" and vice versa as shown in Fig. 4.8B. It seems to become possible only above 440 K, very close to the melting temperature of typical melt-grown crystals. Such type of motion seems also involved in the transition from the polymorph II to III [211–214]. The phase transition is too slow and too close to the melting transition to be followed calorimetrically [203].

A third level of conformational disorder, also frozen-in below the melting temperature, was proposed to be involved in the poling of the PVF2 polymorph II, i.e. the production of a macroscopic dipole of the crystal by application of an electric field

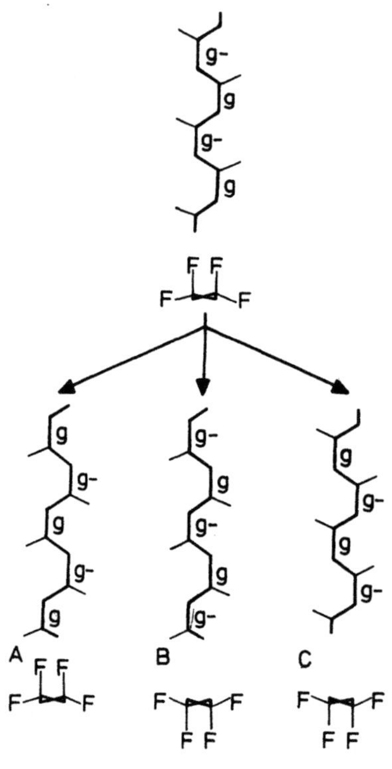

Fig. 4.8. Modes of conformational changes in poly(vinylidene fluoride). "LEFT, UP" to "LEFT, DOWN" transition (A) to "RIGHT, DOWN" transition (B) to "RIGHT, UP" transition (C) Drawn after Ref. [209]

(ferro-electric polymorph II_p). Above about 375 K a polar crystal form can be produced upon application of an electric field. This process is thought to involve the forced conformational change shown in Fig. 4.8C. Thermal motion alone cannot accomplish this conformational change or its reverse below the melting temperature. A direct corollary of this observation is that there is no observable Curie transition for the paraelectric polymorph II in which the poling orientation is thermally unstable, i.e. randomizes after removal of the field.

Ferroelectric polymorphs exist also of the other crystal structures of PVF2 and all have their Curie transition to the paraelectric state above the melting temperature. This could be proven by lowering the melting temperature by copolymerization of vinylidene fluoride with trifluorethylene or tetrafluoroethylene [215–218]. In these copolymers the major condis phase I of PVF2 is stabilized. On poling, the packing of the close to trans-chains becomes more dense and the chains are aligned to produce a macroscopic dipole. The Curie temperature involves the introduction of dynamic conformational disorder through introduction of tg,tg(—) and t_3g conformations at a 15–30 K wide, weakly endothermic transition.

4.4 Polypropylene

Isotactic polypropylene crystallizes below 350 K on quick cooling from the melt increasingly as condis glass [1,219]. This conformationally disordered polypropylene had first been described as being "smectic mesomorphic" [220] or "paracrystalline" [221]. At room temperature this polymorph was shown to persist for over 18 months [222].

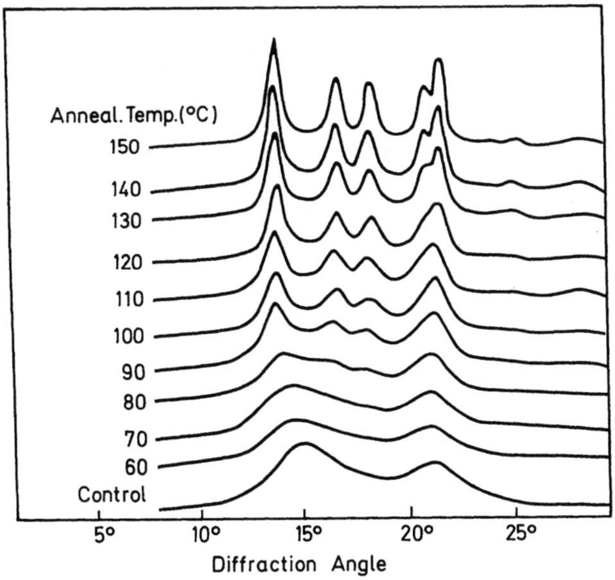

Fig. 4.9. X-ray diffraction pattern of the CD glass of polypropylene (Control) as a function of temperature. The traces suggest a CD glass transition between 340 and 360 K. [Curves courtesy of Dr. W. W. Cox, Research Lab, Hercules, Inc.].

Figure 4.9 illustrates in the lowest curve the poor long-range order of such quenched condis glass. X-ray diffraction, density and infrared analysis as well as ^{13}C solid-state NMR indicate that the "crystallinity" of the condis glass is relatively high (about 40%), the helices have many intramolecular defects (helix reversals) and the intermolecular packing is imperfect [220, 221, 223, 224].

All isotactic vinyl polymers can be arranged in four ways in the usual crystals. Polypropylene has as stable conformation a 2*3/1 helix that consists of successive tg(—) or tg rotational isomeric states (left-handed or right-handed helices). In addition, it was possible for the helices to have the $—CH_3$ groups point "up" or "down", relative to the crystallographic c-axis. In the ideal, monoclinic crystal form I ($P2_1/c$) all four types of helices occur only at specific, symmetry-related positions [47]. The equilibrium melting temperature and entropy are 460.7 K and 15.1 J/(K mol), respectively.

Similar to poly(vinylidene fluoride) (Sect. 4.3) left- and right-handed helices can be inverted by conformational interchanges. Because of the asymmetry of the CHR-group an "up" or "down" helix, however, can not be interchanged within the crystal as in the case of poly(vinylidene fluoride). Melting and recrystallization in the opposite direction is needed. As a result, only one level of conformational disorder, namely that of the helix reversals, can become mobile at a glass transition. Looking for the transition of the condis glass to a condis crystal, it was found, however, that the condis crystal is metastable [221] and reverts on heating with an exotherm of 600 J/mol [225] at about 380 K to the monoclinic stable crystal form I. Because of random "up" and "down" helix placement, the space group is C2/c instead of the full symmetry $P2_1/c$. Figure 4.9 shows the various stages of crystal improvement at various temperatures. Further crystal perfection to correct "up" and "down" mistakes occurs in the melting zone [226–228]. All searches for a possible stable condis phase has been unsuccessful so far. The observed triclinic high pressure phase II [229] and trigonal high temperature phase III [230] seem also not to be conformationally mobile condis phases since they show no fast transition to or from the at atmospheric pressure stable monoclinic crystal phase I [7]. Their long-range X-ray structures are, in addition, much more perfect than that of the condis glass shown in Fig. 4.9.

Since under no conditions there seems to be a stable condis crystal of polypropylene, the reason for the formation of the condis glass must be looked for in the Ostwald Law of successive states which proposes that a new phase will develop step-by-step through successively more stable polymorphs [155]. Efforts to quench the possible crystallization intermediate at temperatures above 380 K failed [219]. Ultra-quenched, glassy polypropylene was shown, however, to crystallize via "nodular, smectic structures" to monoclinic microcrystals [231]. Even at higher temperatures, X-ray diffraction patterns, infrared and NMR evidence were shown to indicate an intermediate "paracrystalline state" of polypropylene [232]. A slow, further development of glassy condis polypropylene to a higher degree of order was also reported [233], suggesting that some perfection can occur just barely above the glass transition of the amorphous polypropylene (270 K).

4.5 Polydienes

Trans-1,4-polybutadiene is perhaps the best-known flexible, linear macromolecule with a dynamic condis phase [1]. The condis phase (crystal phase II) is pseudo-hexagonal [235]. For a long time the disordering transition, extrapolated to equilibrium to occur at 356 K, is recognized as a transition to high conformational mobility [235]. The shear modulus of a typical high-crystallinity sample drops sharply at the transition by about one order of magnitude [236]. Nuclear magnetic resonance proved early high mobility in the high-temperature condis phase [237]. Between 300 and 330 K the second moment of the proton magnetic resonance spectrum drops from 12 to 3 G^2. Figure 4.10 is a thermal analysis curve of the polymer [31], superimposed on such NMR second moment plot [237]. The low temperature heat capacity matches the calculated one from the approximate frequency spectrum of the rigid crystal and glass (see also Fig. 1.3). The low temperature crystal from I is monoclinic ($P2_1a$) and shows a disordering transition entropy to the condis phase of $\Delta S_d = 21.9$ J/(K mol). The final entropy of transition from the condis to the isotropic state is $\Delta S_i = 8.6$ J/(K mol) [238]. The total entropy of fusion is thus 31.5 J/(K mol), very close to that of cis-1,4-polybutadiene [$\Delta S_i = 32.3$ J/(K mol)][31]. This second geometrical isomer of 1,4-polybutadiene does not show any condis phase [31]. For both isomers one can easily identify three flexible bonds, i.e. according to the "fusion rules" of Sect. 1.1, one expects 3×9.5 J/(K mol) for the overall entropy of fusion of either isomer, as is observed. The low-temperature transition of the

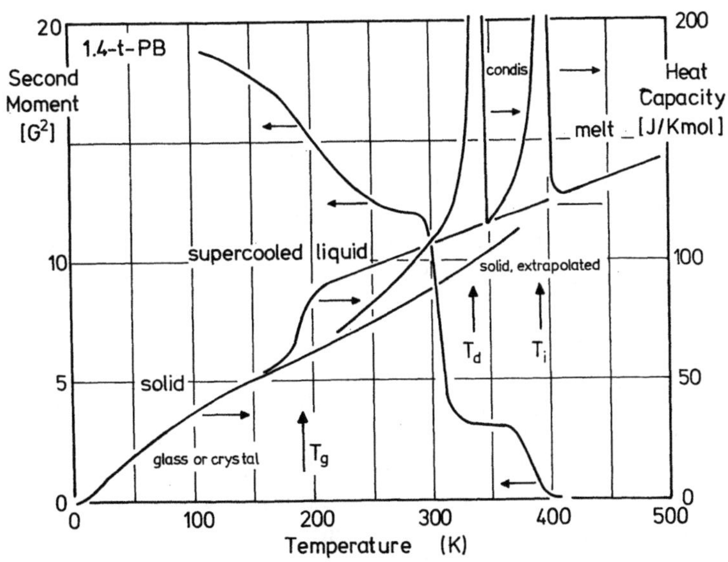

Fig. 4.10. Thermal analysis and proton NMR second moment curves of trans-1,4-polybutadiene. The heat capacity for the solid (glass and low temperature crystal form I) have been computed by fitting at low temperature to an approximate skeletal vibration spectrum. The gradual decrease of the second moment of the proton resonance spectrum below T_d comes mainly from increasing mobility in the amorphous phase, Ref. [237]. The melting curve is of an approximately 90% crystalline sample, Ref. [31]. Compare also to Fig. 1.3

trans-isomer seems thus to permit dynamic conformational disordering about two of the three bonds. From symmetry considerations alone, one can conclude that intramolecular rotation about the =C—C-bonds with symmetry number 6 must be much easier than about the —C—C-bond with symmetry number 3 [47].

More detail about this mobility and disorder in the condis crystal comes from the interpretation of ^{13}C MAS NMR spectra and from deuterium NMR on a polymer deuterated at the methylene groups [236]. The ^{13}C-spectrum of the liquid state shows a chemical shift of 130 ppm for the methine carbons and 32.8 ppm for the methylene carbons (TMS = 0 ppm). On crystallization to the condis phase the methylene carbon signal shifts downfield by 0.6 ppm, while the methine signal shows no change. Lowering the temperature below T_d, additional signals occur for both, the methylene carbon (at 35.2 ppm) and the methine carbon (at 131.2 ppm). These signals can be assigned to the monoclinic polymorph. On fast cooling to low temperature (318 K) all three methylene signals could be observed simultaneously, indicating a three-phase sample with stable monoclinic crystals. metastable condis crystals and melt. The metastable phases must change to the corresponding glasses at sufficiently low temperature (see Fig. 1.1), and indeed, their NMR signals broaden significantly below 320 K. The upfield shift in going from the monoclinic methylene (35.2 ppm) to the condis methylene (—1.8 ppm) and ultimately to the melt methylene (—2.4 ppm) can be used to estimate the conformational disorder to be about 75% of that of the melt, in close accord with the entropy measurement. The deuterium spectrum of the methylene groups in the condis state is narrowed to a degree that is indicative of diffusive rotation and conformational interconversions with correlation times of less than 10^{-8} s [236].

The molecular chain conformation in the crystal form I is known from the X-ray structure [239]. The bonds adjacent to the double bond are rotated out of the trans-conformation by +71° and —71°. This leaves the central CH_2—CH_2 group in the low-energy, trans-conformation. On transition to the condis phase the chain position and orientation remains unchanged, but the chain separation increases by 7.5% and the length decreases by 4% with an overall volume change of +9% [234, 235, 240]. The entropy increase due to the volume increase alone can be computed from expansivity and compressibility to be about 3.2 J/(K mol) [241], leaving about 18.7 J/(K mol) for conformational disordering. In the rotational isomer model this would correspond to a transition from one to 9.5 conformers per repeating unit. Molecular mechanics computations have resulted in two to nine low energy rotational isomers per repeating units [240, 241].

Although crystallization of condis crystals from the melt leads to larger fold-lengths (15–25 nm) than crystallization of form I crystals (8–10 nm), there is no indication of extended chain crystallization [238]. Mechanical deformation (drawing) of the condis phase leads, however, to considerably higher chain extension and corresponding higher modulus than drawing of polymorph I [242]. Diffusion of disorder along the chain and through the folds, as observed for polyethylene and polytetrafluoroethylene, seems thus more difficult for trans-1,4-polybutadiene. Once crystallized as form I, one observes on transition to the condis phase an abrupt doubling of the fold-length. Such doubling of fold-length on annealing has also been observed for other macromolecules of intermediate mobility such as aliphatic nylons and polyethylene in their low-temperature crystal forms [155]. A detailed defect mechanism for the unfolding to twice the thickness without large scale chain motion has been proposed, and seems to have general applicability [243].

On slow cooling (10 K/min or less) the trans-1,4-polybutadiene crystallizes first as condis crystals and then converts at lower temperature completely into the ordered crystal phase I. Both transitions have about equal 20 K supercoolings and can easily be followed by thermal analysis [31]. On fast cooling, the condis crystals convert only incompletely to crystal form I, i.e. the low temperature polymer sample represents a three-phase structure consisting of crystal phase I, supercooled condis crystals or CD glass, and supercooled melt or glass. This three phase structure was also proven by NMR [236] (see above). A typical sample cooled at about 300 K/min has at low temperatures a composition of 59, 9, and 32 wt.-% of phase I crystals, condis crystals, and amorphous phases, respectively [31].

A separation of the broad glass transition centered about 190 K into one portion due to the amorphous glass and one to the CD-glass was not possible, indicative of small differences in T_g. From the similar heat capacity of melt and condis crystals shown in Fig. 4.10, an identical ΔC_p is expected for amorphous and CD glass [31]. An interesting observation was that poorly crystallized samples increase more (about twice as much) in condis crystallity than in form I crystallinity when annealed below T_d. This seems to indicate that, as in polypropylene (see Sect. 4.4), rigid crystals develop only over the path of the condis phase, even though direct transformation would be thermodynamically possible. Annealing a monoclinic sample above T_d, i.e. after transition to the condis phase, did not change the amount of condis crystallinity [31].

The crystallization kinetics from the melt to condis crystals and then from condis crystals to form I crystals shows two-stage crystallizations and exceptionally low Avrami coefficients [244] (0.02–1.74). Little is know about the detailed mechanisms, except that the kinetics indicates significant differences from "normal" crystallizations.

The closely related *trans-1,4-poly(2-methylbutadiene)* (gutta percha) has also two crystalline polymorphs. The stable, monoclinic α-crystal form ($P2_1c$) grows from the melt above 318 K (T_m = 353 K) and has an entropy of fusion of 36.4 J/(K mol), indicative of full conformational order. The second polymorph, the orthorhombic β-crystal form (Pnam) grows at lower crystallization temperatures, but has only a somewhat lower entropy of fusion [29.7 J/(K mol)]. Conversion from β to α is not possible, or at least extremely slow. Both polymorphs have been asigned distinct chain conformations [245], i.e. neither seems to show large scale dynamic conformational disorder.

It was shown already above that *cis-1,4-polybutadiene* melts in one step with the expected entropy of fusion. In contrast, *cis-1,4-poly(2-methylbutadiene)* (natural rubber) has a more complicated fusion and crystallization behavior [47, 155]. The reported entropy of fusion of the common monoclinic ($P2_1a$) crystal polymorph is only 14.4 J/(K mol), less than half of the expected value [7]. The crystal structure has been reported statistically disordered, but only relative to packing of chains that are mirror images of each other along the crystallographic a-axis [246]. Such geometric disorder cannot account for a 50% decrease in entropy of fusion. A full study of the thermodynamic functions as available for the polybutadienes would be of value.

At elevated pressure a hexagonal polymorph of cis-1,4-poly(2-methylbutadiene) has been reported. A typical crystallization condition is 300 K temperature and 300 MPa pressure [247]. It was suggested that this high-pressure polymorph derives from the other polymorphs through introduction of interchain packing defects. The chain-axis repeat was found to be unchanged. The general behavior was compared to the hexagonal condis phase of polyethylene [248]. In analogy to the early descriptions of polyethylene, cis-1,4-poly(2-methylbutadiene) was described as a "rotor phase solid" or being similar to a "smectic liquid crystal". Overall this phase may, perhaps better be classified as a condis crystal although full proof is still lacking. As in polyethylene, tapered, large lamellae with secondary thickening were observed, indicative of larger mobility in the condis phase (see Fig. 4.3). The ultimate thickness of the lamellae was, however, much less than found in the extended chain crystals of polyethylene. Quenching to a CD glass was possible by cooling under pressure, but a conversion to the more stable, atmospheric-pressure polymorph was also possible at elevated pressure. It seems thus that the condis crystal phase of cis-1,4-poly(2-methybutadiene) is metastable because of nucleation difficulties of the more stable phase [248]. Perhaps cis-1,4-poly(2-methylbutadiene) should be compared to polypropylene (see Sect. 4.4) with a disordered intermediate before ultimate crystallization. Because of the larger mobility about the =C—C-bond, the condis phase can be observed at higher temperature (and pressure), but full ordering seems not to be possible, even in the low pressure polymorph.

4.6 Other Flexible Carbon-Backbone Macromolecules

Besides the eight well documented condis phases of flexible, carbon-backbone macromolecules described in Sects. 4.1 to 4.5, there is evidence of an even larger abundance of condis crystals as stable phase, intermediate in crystallization, and possibly also on large scale deformation. In this section some indication of the existence of condis phases for aliphatic nylons, polyallene and poly[(S)-3-methylpentene-1] will be reviewed. Lateral and longitudinal disorder was noted much earlier for a larger series of polymers [249], but little is known about the dynamic nature, i.e. if these macromolecules must be considered CD glasses or if they are stable condis crystals at any temperature.

Nylon 6, *nylon 11*, and *nylon 12* have been reported to grow as extended chain crystals when crystallized at pressures of up to 800 MPA [250–252]. Based on the needed high mobility in the crystalline state for chain extension, it is quite likely that under these conditions the polyamides possess the conformational mobility and disorder of a condis crystal. No new crystalline polymorph was, however, discovered at elevated pressure. All three nylons result on cooling and pressure release in α-crystals. This applies even to nylon 12 that grows at atmospheric pressure as the β-polymorph. Heating of the γ-polymorph of the nylons at elevated pressure results in a quick conversion to the α-form with an easily observable exotherm [253]. At atmospheric pressure this transition is not possible, or so slow that it is hardly detectable by calorimetry [7] and needs treatment with phenol to mobilize the H-bonds for the conformational readjustment [7,47]. Rather larger mobility is also indicated by a high concentration of non-hydrogen-bonded NH-groups detected by infrared analysis [250]. All these observations suggest that the condis phases of the nylons may develop gradually without major transition entropy. Besides conformational mobility there is, in addition, also the possibility of chemical bond rearrangement through trans-amidation [250]. Without the restraint due to amide groups and H-bonds, the CH_2-sequences find themselves at the temperatures in question in a much higher thermal energy environment than in crystalline polyethylene and conformational mobility should be likely. The polyethylene equilibrium melting temperature is 414.6 K, while nylon 6 melts, for example, at 533 K, a temperature at which polyethylene is in the condis phase even at 500 MPa presure (see Fig. 4.1).

Polyallene with a chemical structure represented by

$$(CH_2{-}\underset{\underset{\displaystyle CH_2}{\|}}{C}{-})_n \qquad (4.3)$$

has a common 2*2/1 helix structure for three crystal polymorphs. One of these polymorphs, form III, was reported to have a disordered "paracrystalline" state [254]. All X-ray layer lines are nearly streak-like, exhibiting no sharp spots, but indicating the same helix repeat as the ordered polymorphs. This diffraction pattern suggests a condis state without correlation between left and right-handed helices. The disorder in this crystal is, however, frozen over the analyzed temperature range, i.e. the analyzed form III should be classified as a CD glass (see Fig. 1.1). The metastable nature of the glass is proven by a phase transition to crystal form II on annealing. The crystal form II still contains random mixtures of right-handed and left-handed helices, but these have now a laterally ordered packing (space group $P2_1/m$, monoclinic). Only slow cooling from the melt produces the stable, orthorhombic crystal polymorph I (Pnan) with matching left- and right-handed helices packing in the same unit cell. The form II is grown from the melt on quenching. From the few data available it seems that polyallene has a crystalline behavior similar to that of polypropylene (see Sect. 4.4).

Another interesting case of conformational disorder is provided by isotactic *poly[(S)-3-methylpentene-1]* (P3MP1)

$$(CH_2{-}\underset{\underset{\displaystyle H\underset{\underset{\displaystyle C_2H_5}{|}}{C}{-}CH_3}{|}}{C}H{-})_n \qquad (4.4)$$

In this case the side-groups may take-on statistical disorder with respect to rotation of the terminal —CH—CH_3. The two minimum-energy conformational isomers of different side-group orientation seem to have only small energy differences when packed in their tetrahedral crystal (space group $I4_1$) [225]. The backbone helix is always of the same type (left-handed 2*4/1). There has been no report as yet if this disorder is dynamic, and in case it is, whether it freezes via a first order transition or a glass transition.

4.7 Polydialkylsiloxanes

Polydialkylsiloxanes represent a group of flexible macromolecules that have not only interesting low and high temperature materials properties (silicones), but they may also show special electronic properties. The simplest member of the homologous series is *polydimethylsiloxane*. It exhibits no condis phase and seems to have "normal" thermodynamic properties [256–257]. On samples of crystallinity varying from zero to 0.90 the glass transition at 148 K was found to have at zero crystallinity a somewhat high change in heat capacity of 27 J/(K mol), and the melting transition, extrapolated to 1.00 crystallinity, at 219 K a somewhat low entropy of fusion of 11.8 J/(K mol). This trend is continued for *polydiethylsiloxane* with a change in heat capacity of 30.7 J/(K mol) at a glass transition of 135 K and a combined entropy of fusion of 15.7 J/(K mol) [258]. A disordering transition is observed at 206 K [ΔS_d = 8.5 J/(K mol)] and a change into an almost isotropic phase at 276 K. A small amount of residual order that is distinctly different from a liquid crystalline state exists between 276 and perhaps 320 K [ΔS_{i_2} = 0.9 J/(K mol)] [259]. It was initially interpreted as a "viscous crystalline" phase [260] and shows a sharp equatorial X-ray diffraction line at 0.8 nm and an endotherm that changes its peak temperature depending on the thermal history [261]. Of interest to the present discussion is, however, the crystal phase between 206 and 276 K that is clearly a condis phase [1]. Computations of the conformational energy minima of isolated chains show the possibility of two conformational isomers, mainly caused by steric interactions by the side chains [262]. The motion in the condis crystalline and amorphous regions in polydiethylsiloxane is similar as shown by proton nuclear spin-spin relaxation-time measurement [263]. Solid state ^{13}C and ^{29}Si MAS NMR on the crystals of polydiethylsiloxane [95,131] revealed the possibility of a continuous change of conformation in the crystalline state between T_m and 190 K. A doublet ^{13}C absorption-line, indicative of two different —CH_3 environments, changes to a singlet. This may be an indication that similar to the cyclosiloxanes (see Sect. 3.3) intermolecular energy considerations play a more important role in the determination of chain conformations of siloxanes than in carbon-chains where intramolecular energies are dominant.

A similar behavior as in polydiethylsiloxane was also observed for *polydipropylsiloxane* [264]. The transitions are now at T_g = 164 K, T_d = 218 K, T_{i_1} = 333 K, T_{i_2} = 479 K with the corresponding entropies for the best crystallized materials of 9.0, 8.6 and 0.8 J/(K mol), respectively, for the first order transitions. Between 218 and 333 K the polymer is most certainly again a condis phase. One must note that the complete polymorphism develops in this polymer only with higher molecular mass and that the transition temperatures increase in going from polydimethylsiloxane to polydipropylsiloxane almost linearly. This contrasts the hydrocarbon analogs with one side-chain per repeating unit where the melting temperature decreases in going from polypropylene to poly-1-pentene from 461 to 403 K. One may take this as another indication of the major influence of the side-chain on the backbone conformation.

Quite recently *polysilanes* with longer hydrocarbon side chains have been made and analyzed by thermal analysis, X-ray diffraction and spectroscopy [265]. The melting temperatures of the di-n-hexyl, -heptyl and -octyl polymers were found to be about

310–320 K and, as in the siloxanes, the main chain conformation (extended, planar zig-zag) seems to be governed by side-chain crystallization.

4.8 Polyphosphazenes

Alkoxy and aryloxy polyphosphazenes with the general structure

$$-\underset{OR}{\overset{OR}{\underset{|}{\overset{|}{P}}}}=N- \qquad (4.5)$$

are a group of macromolecules of which many have stable mesophases over wide temperature ranges. In a general review 16 of such polymers have been described. It was suggested that all of these mesophases are condis crystals [1]. The polymers are flexible with small, intrinsic barriers to rotation, as indicated by their low glass transition temperatures that range from about room temperature down to 180 K depending on substitution [267]. In the fully ordered, low-temperature polymorph the chain conformation of the polyphosphazenes is cis-trans planar with a fiber repeat length of 0.48 to 0.49 nm. The thermal behavior is dominated by two endothermic, first-order transitions that limit the condis state.

The disordering transition of *poly[bis(trifluoroethoxy)phosphazene]* [R:CF_3 —CH_2—, T_d = 365 K] shows an entropy change ΔS_d of 34 J/(K mol) that represents almost all of the entropy change observed in transitions. On changing to the isotropic phase at 513 K one finds only an additional 2.3 J/(K mol) [268]. Each of the two transitions is accompanied by an almost equal 6–7% volume increase. Such large volume change is normally connected with a much larger entropy change than observed. In any case, there is little additional conformational disordering possible at T_i [269]. Solid state ^{19}F, ^{1}H and ^{31}P NMR line-width studies (CW) and X-ray analysis showed also clearly that at T_d the orthorhombic low-temperature crystal goes to a dynamically disordered pseudohexagonal condis crystal with rapid, large-amplitude side-chain and main-chain motion [270]. On growth from dilute solution chain-folded, monoclinic lamellae can be grown, while from somewhat more concentrated solution chain-folded, orthorhombic, fibrillar spherulites [271] in form of films could be cast. As grown, or after annealing at temperatures below T_d, the fold period was 15.6 nm for the single crystals and as much as 25 nm for the spherulites (SAXS). On heating through T_d, or on direct crystallization into the condis state, irreversible chain extension occurs, as was also observed for polyethylene or polytetrafluoroethylene (see Sects. 4.1 and 4.2). Fracture surfaces now are similar to the characteristic for the stiff-chain, practically fully-extended structures of poly(p-phenylene terephthalamide) (see Sect. 6.4) [272]. On cooling through T_d the chains remain extended and a more stable orthorhombic crystal polymorph (γ) is reached.

Similar polymorphism and enhanced ordering on heating above T_d was also observed and is expected for a group of three related phosphazenes. *Poly[bis(phenoxy)phosphazene]* (R:C_6H_5—, T_d = 423 K) was similarly analyzed with respect to X-ray diffraction and electron microscopy as the trifluoroethoxy compound [273, 274].

Poly[bis(p-methylphenoxy)-phosphazene] (R:CH_3—C_6H_4—, T_d about 430 K) was cooled and heated through T_d while continuously recording wide- and small-angle X-ray diffraction patterns (synchrotron radiation)[275]. Again, crystal-improvement, chain-extension and increase in crystallinity was observed on cycling through T_d. Finally some thermodynamic data are also available for the closely related *poly[bis(p-chlorophenoxy)phosphazene]* [R:Cl—C_6H_4—, T_d = 442 K, ΔS_d = 27 J/(K mol)][269].

4.9 Summary and Conclusions

The discussion of the behavior of condis crystals of flexible linear macromolecules and some of the homologous oligomers covers a wide range of conformational disorder-effects and accompanying motion. For many of the macromolecules in the condis phase chain-extension after crystallization with chain folding is possible [polyethylene, polytetrafluoroethylene, poly(vinylidene fluoride), polychlorotrifluoroethylene, some aliphatic nylons and polyphosphazenes]. Such extended chain crystals are exceptionally stable and close to equlibrium.

For polyethylene and polytetrafluoroethylene the condis phase could be traced to the oligomeric homologues with special effects due to chain ends (paraffins) and whole molecule rotations (perfluoroalkanes). While the plastic crystal phase in cycloalkenes (Sect. 3.2) and substituted benzenes (Sect. 5.1.1) is restricted to dynamic disorder of a single conformer about rotation axes normal to the planes of the molecules, perfluorobutane and perfluorohexane have dynamic disorder restricted to motion about axes parallel to the molecular axis.

Several of the described condis phases are CD glasses (Sect. 4.2–4 and 4.6) with different paths into the glassy states. Some of these CD glasses seem to have no corresponding, stable condis phase, but are produced as intermediates in the path from melt to stable crystal [polypropylene]. The ferroelectric effect of poly(vinylidene fluoride), polytrifluoroethylene and its copolymers seems connected to condis glass and disorder transitions (Curie temperatures). Condis states seem also to have enhanced chain mobility on mechanical deformation.

For the aliphatic nylons there is an indication of a non-cooperative introduction of conformational disorder. For the polyphosphazenes the transition from the condis phase to the isotropic melt has an exceptionally small entropy gain while retaining normal volume changes. Both observations are still in need of further investigation.

The longer side-chain siloxanes display, besides a seemigly normal condis phase, residual order in the melt that is different from liquid-crystalline order. As seen for cyclosiloxanes in Sect. 3.3, steric hindrance of side-chains and intermolecular packing effects seem to have larger influences on siloxanes and also on polyphosphazenes.

The condis state of flexible macromolecules and oligomers is thus central to the understanding of such important observations as chain extension to produce high modulus, perfect, equilibrium crystals response to mechanical deformation; production of intermediates states on crystallization; and ferroelectric effects. In Section 5 it will now be analyzed what changes in the condis state are possible through the introduction of rigid segments into flexible small and large molecules.

5 Condis Crystals and Their Relation to Liquid Crystals

The common structural characteristics of thermotropic liquid crystals is a rigid rod- or disc-like mesogen of sufficient size [3]. This enables the molecules in the liquid state to retain a limited degree of orientational order that is lost only above the temperature of isotropization T_i. Analysis of a number of nematic liquid crystals of small molecules and macromolecules with the mesogen coupled via a flexible spacer to the main chain showed that the orientation entropy ΔS_i is only 1–3 J/(K mol), i.e. less than 10% of the entropy of full orientation of the mesogen alone (see Sect. 1) [1]. For nematic macromolecules with mesogens in the main chain separated by flexible spacers, ΔS_i increases to 2–30 J/(K mol), but the increase in ΔS is to some degree due to some ordering of the flexible spacers, rather than a better orientation of the mesogens [276].

In this Section rigid mesogens are built-up first by linking phenylene groups (Sect. 5.1). As in the case of the cycloalkanes, the various motional processes in the solid state are followed from the first member of the homologous series, benzene, to the macromolecule, poly-p-phenylene. One finds regions of motion involving jumping between positions of identical symmetry, condis mesophases, and liquid crystals, as well as rigid crystals. The equilibrium state of matter depends on chain-length and temperature. In Sect. 5.2 several typical low molecular mass liquid crystals are discussed that maintain conformational disorder on cooling, i.e. show condis crystal phases before going into the fully ordered crystal state. Combinations of rigid mesogens with flexible spacers are then discussed in Sect. 5.3. Reduction of the flexible spacer length must lead more and more to a loss of the positional mobility (translation) required for a liquid crystal and may give states intermediate between liquid crystal and condis crystal. The limited knowledge about these increasingly stiff molecules is discussed in Sect. 6.

5.1 Poly-p-Phenylenes

As in the case of cycloalkanes, the types of motion in the homologous series of phenylenes is strongly size dependent. Figure 5.1 shows the thermal transition temperatures as a function of phenylene groups and in Table 5.1 the transition entropies are listed. The first member of the series shows no thermal transitions below the melting temperature (Sect. 5.1.1). With the second and third, a condis phase becomes obvious (Sect. 5.1.2) and with the fifth, liquid crystalline states exist in addition (Sect. 5.1.3). Poly-p-phenylene (Sect. 5.1.4) is the limiting member of these molecules.

5.1.1 Benzene and Related Compounds

Heat capacity measurements of *benzene* and *naphthalene* have been available for many years [277]. No remarkable features were observed from 90 K up to the melting temperature. Figure 5.2 shows the heat capacity of benzene superimposed on the proton second moment of the NMR spectrum [278]. At 90 to 120 K the second moment narrows from the value expected for the rigid lattice (calculated 9.62 G^2, measured 9.72 G^2) to that of a crystal permitting jump-reorientation about the six-fold

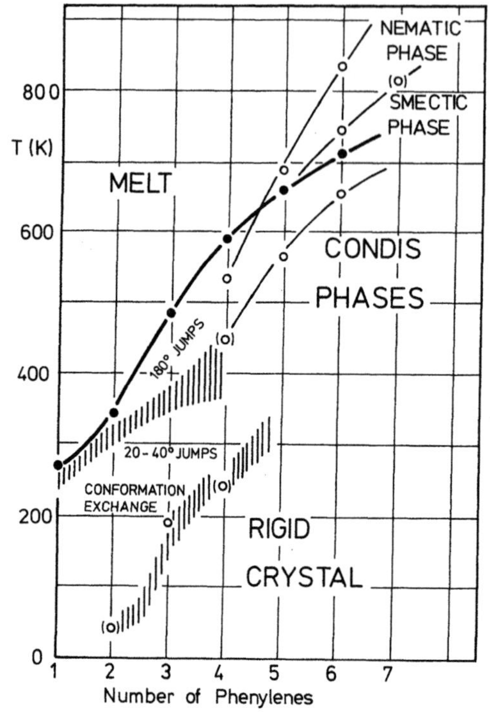

Fig. 5.1. Transition temperatures of polyparaphenylenes. Filled circles: large entropy of transition (see Table 5.1). Open circles: low entropy of transition. In parentheses: weak or questionable transitions by thermal analysis. Shaded area: changes in motion indicated by NMR

molecular axis (calculated 1.68 G^2, measured 1.61 G^2). The activation energy for this process was found from the spin-lattice relaxation time [279] to be 17.5 kJ/mol and led to a frequency of about 10^4 Hz at 90 K, increasing to 10^{11} Hz just below the melting temperature of 278.6 K [278]. The calculations brought quantitative agreement between the model of uncorrelated jumping between various symmetry

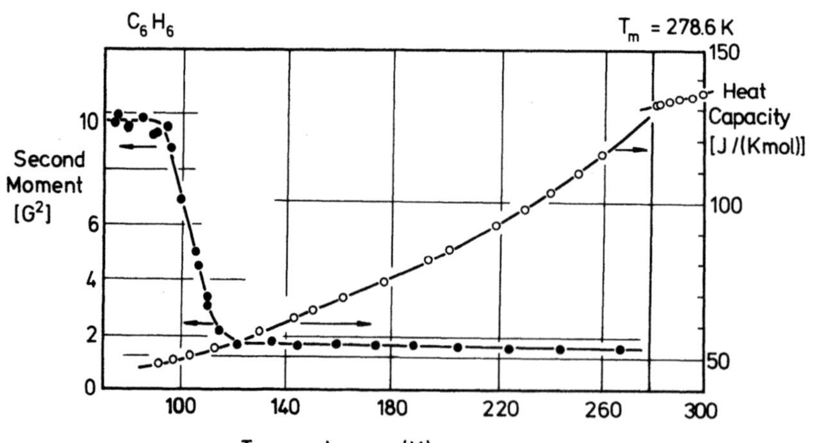

Fig. 5.2. Heat capacity and second moment of the proton NMR spectrum of benzene. Drawn after Refs. [277,278], respectively

Table 5.1. Transition temperatures and entropies for p-polyphenylenes

	T_d (K)	S_d [J/(K mol)]	T_i (K)	S_i [J/(K mol)]
Benzene	no	—	278.6	35.3
Biphenyl	(40)	0.0	344.1	54.6
Terphenyl	193	0.49	486	72.8
Quaterphenyl	(243)	0.0		
	(450)	?	593	63.7
Quinquephenyl	(565)	?		
	661	63.8	691	1.33
Sexiphenyl	663	?		
	713	39.41		
	748	7.35	838	2.74
Septiphenyl[a]	818	?		
Polyphenyl	—	—	>1000	?

[a] M. Busch and W. Weber, J. Prakt. Chem. *146*, 1 (1936), reported as melting temperature, but probably transition to the liquid crystalline state

related positions and structural data [279]. The RMS amplitudes of vibration of the molecule in the molecular plane were calculated from X-ray and neutron scattering to be 2.5°, 4.9° and 7.9° at 138, 218, and 270 K, respectively [280,281]. About 40 K below the melting temperature there is some suggestion of a beginning slow reorientation of the benzene molecule about an axis within the plane of the molecule with an activation energy of 70 kJ/mol [282]. Despite the smooth heat capacity below the melting temperature, a large degree of mobility exists thus in benzene.

Naphthalene, in contrast to benzene, did not show any NMR-spectra line-width narrowing up to its melting temperature of 353 K. The mean experimental second moment was 9.1 G^2 compared to 10.1 G^2, estimated for the rigid crystal. Measurement of spin-lattice relaxation times indicated, however, also a slow reorientational jump motion about an axis normal to the molecular axis [283]. An activation energy of 105 kJ/mol was derived. Molecular dynamics simulations suggest that this reorientation about the axis of greatest inertia occurs with a frequency of 100 MHz within 20 K of fusion (353.6 K) [284]. Still, no plastic crystal behavior as found in cyclohexane and related compounds (see Sect. 3.1.1) is indicated for benzene or naphthalane, even close to the melting temperature.

Furan C_4H_4O is a cyclic molecule that is also almost planar [285], but lacks the symmetry of benzene. Calorimetrically one finds a disordering transition (T_d = 150.0 K) with smaller entropy change [ΔS_d = 13.6 J/(K mol)] before fusion (triple point at 187.6 K) with larger entropy change [ΔS_i = 20.3 J/(K mol)] than in benzene [286]. The second moment of the proton NMR spectrum drops from the rigid value 7.5 G^2 (estimated 5 ± 2.5 G^2) to the one for reorientation within the molecular plane (2.75 G^2) at precisely T_d [287], i.e. while benzene can undergo jump-reorientation without entropy change at a similar temperature, furan must absorb a considerable entropy due to its lack of symmetry for such motion. The crystal structure of the high-temperature mesophase is orthorhombic (C_{mca}) and permits statistical placement of each furan molecule in one of four coplanar positions so that one H-atom of a furan ring fits into the indentation of the center of the neighboring furan ring that is arranged at right angles [285]. Four disorder positions should give an entropy contribution of R ln 4 = 11.5 J/(K mol), in good agreement with the calorimetry [286]. The high temperature crystal-form is thus not a 3-dimensionally reorienting plastic crystal, but shows jump-like motion about an axis normal to the molecular plane, similar to crystalline benzene. On transition to the isotropic phase, ΔS_i of furan must thus be more than expected for a plastic crystal [20.3 instead of 7–14 J/(K mol)]. On cooling to the rigid crystal phase, the furan crystal becomes tetragonal without much change of the molecular position, but with ordering of the oxygen locations.

Similar reorientations within the plane of the cyclic molecules have been discussed for *pentafluorochlorobenzene* (ΔS_d = R ln 6) [288], and other *asymmetrically hexasubstituted benzenes* [289]. In several of the latter molecules $-CH_3$ and —Cl groups are used as substituents. These have almost equal molecular volume and there is a good indication of gradual unfreezing of the reorientation above 120 K, but it is not clear if the entropy gain is also gradual, or if the crystal at low temperature has a frozen, random orientation within the planes of the molecules.

Hexamethylbenzene itself shows, because of its high symmetry, again a narrowing of the second moment of the proton NMR spectrum independent of thermodynamic phase transitions [290,291]. In the triclinic low temperature phase III, up to 116 K, one finds only increasing rotation of the $-CH_3$ groups. In the triclinic intermediate phase II (116–383 K) reorientation about the C_6-axis in the plane of the molecule becomes possible. The full reduction in the second moment of the proton NMR spectrum for this process occurs between 150 and 200 K and not at the transitions. A further phase change at 383 to the orthorhombic (polymorph I) shows a volume expansion, but no further change in the second moment. The reorientation of the molecule remains thus in the plane of the molecule. Fusion takes place at 450 K with gaining of full disorder.

5.1.2 Biphenyl and Terphenyl

Biphenyl is the first molecule of this series that may show rotation about the bond connecting the phenyl rings. On rotation out of the planar conformation the overall molecule remains linear because of the sp^2-bond-geometry, but it loses its center of inversion. In the gas [292] and liquid [293] phases the ortho-hydrogen repulsion is sufficient to yield 42° and 34° twisted conformations, respectively. Surprisingly X-ray diffraction shows a seemingly planar molecule (monoclinic, $P2_1/a$, two molecules per unit cell) [294,295]. Heat capacity measurements between 10 K and room temperature showed no evidence of heats of transition (precision limit about 10 J/mol) [277,296]. The libration tensor determined from X-ray data at room temperature [297] and 110 K [298] revealed, however, an unusually large libration amplitude about the long axis of the molecule (root mean square libration amplitudes 6.8° and 10.4° at 110 and 293 K, respectively). Neutron scattering on $C_{12}D_{10}$ showed that below 40 K superlattice reflections appear, corresponding to a doubling of the unit cell in the b-direction [299]. Neighboring biphenyls in the b-direction are now twisted in opposite directions with a twist-angle of about 10°. This loss of the center of inversion of the molecule was found first by Raman and infrared spectroscopy [300]. On cooling, new bands corresponding to this loss of the center of inversion appear gradually and reversibly below 75 K in the phonon and intramolecular frequency-range without changes in the high temperature spectrum. Because of the gradual

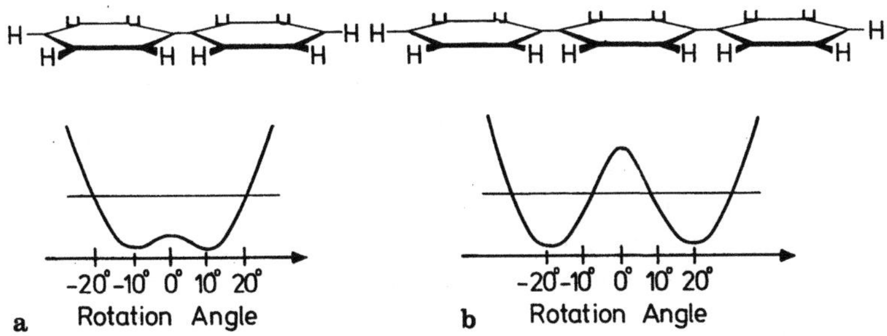

Fig. 5.3a and b. Schematic drawing of the potential energy of biphenyl and p-terphenyl as a function of the torsional angle

change, the 40 K "transition temperature" in Fig. 5.1 is placed in parentheses. The double-well potential that keeps the gaseous biphenyl at a torsion angle of 42° must be modified by an inter-molecular potential to a shallow, double-well potential as depicted in Fig. 5.3a [301]. At room temperature there is, in addition, evidence of a slow, large-amplitude (180°) jump reorientation with an activation energy of about 95 kJ/mol [283].

While in biphenyl the transition to conformational disorder in the crystal is gradual (displacive transition), *terphenyl* shows a clear order-disorder transition in changing into the condis crystal [301]. The double-well potential is shown in Fig. 5.3b. The heat capacity measurement of terphenyl (10–300 K) shows a weak endotherm spread over 20 K with a heat of transition of 86 J/mol centered at 193 K [302]. It is impossible for the molecule to have full, random access to the two conformational positions since in this case one would expect a conformational entropy change of R ln 2 or 5.76 J/(K mol). Indeed, appreciable orientational correlation of neighboring molecule exists up to at least 250 K as proven by incoherent neutron scattering [303]. The low temperature crystal-form is triclinic ($P\bar{1}$) [304], but closely related to the high-temperature, condis crystal (monoclinic $P2_1/a$, similar to biphenyl) [305] with double the unit cell dimensions in both the a and b dimensions. The molecules are aligned in both crystal forms along the c-axis that sees no change during the transition. In the high-temperature crystal form, the molecules appear as a time-average flat, but with large libration angles of the central phenylene (root mean square amplitudes of 16.1° at 300 K and 15.7° at 200 K) [306]. In the low-temperature crystal-form the outer two rings remain almost coplanar (5° displacement from the mean) and the central ring is located in one of the two potential energy wells indicated in Fig. 5.3b (16° from the mean, opposite to the displacement of the outer rings) [307]. The larger displacement angle in p-terphenyl can be accounted for by the doubling of the interfering o-hydrogens when compared with the biphenyl. The potential energy barrier between the two conformational isomers is about 4.4 kJ/mol [303]. The second moment of the proton NMR spectrum [304] stays constant from 250 to 350 K at $8.20 \pm 0.3\ G^2$, a value that corresponds to a crystal with uncorrelated conformational disorder between the two minima indicated in Fig. 5.3b. Above 350 K, the second moment drops to a plateau of $3.80\ G^2$ which remains then up to the melting temperature (486 K). These higher temperature second moments coupled with spin-lattice relaxation data indicate an additional unfreezing of, first,

180° jumps of the two external phenyl groups and then of the central phenylene [308]. The activation energy is 80 kJ/mol and the motions are assumed to be uncorrelated. Although there may also be an anomaly in the heat capacity in the region of the beginning 180° jump motion [309], overall, the development of increasing dynamic disorder in the condis phase is not mirrored by an significant heat or entropy of transition. The ultimate transition into the molten state at 486 K is accompanied by an exceptionally large transition entropy [310] (see Table 5.1), indicative of more than gain of just translational and orientational entropy.

5.1.3 Quaterphenyl and Higher Oligomers

With the p-terphenyl (Sect. 5.1.2) all motions in the condis phase are already illustrated. Analogously one finds for the *quaterphenyl* at 243 K a change from a triclinic ($P\bar{1}$) low temperature crystal structure with the outer phenyl groups rotated ±58° out of the mean plane and, at the same time, the inner phenyl groups rotated +11.3° from the mean plane with alternating rings displaced in opposite directons [311]. Figure 5.4 shows the conformation. The monoclinic high temperature crystal form ($P2_1/a$) [312], the condis crystals, appear again flat from X-ray diffraction results because of exchanging between two limiting conformations. The libration tensor at 300 K gives root mean square amplitudes of 13.4° for the two inner rings and 8.5° for the outer [312]. The transition at 243 K may again be with little or no entropy effect [313,314]. Proton NMR spectra show a line-width narrowing after a plateau from 250 to 350 K. The second moment changes, similar to that of the p-terphenyl, from 8.30 G^2 to 3.90 G^2. The two levels of the second moment agree with an uncorrelated jumping between the two possible conformational isomers in the condis state below 350 K and an additional, two-stage slow 180° jumping of the outer and inner rings with activation energies of 80 and 130 kJ/mol, respectively [308]. At the upper end of the narrowing of the proton NMR line width one finds a thermal anomaly [about 450–540 K, 2 J/(K mol) entropy] [309]. From phase diagrams of solutions with higher polyparaphenyls one can extrapolate that a metastable nematic liquid crystal would have a transition to the isotropic state at about 533 K [315]. The rigid rod-like structure of sufficiently long paraphenyls permits thus orientational order in the melt (see Table 5.1 and Fig. 5.1). Both attraction between the chains and the rod-like geometry contribute to the stability of the nematic state [316].

While in p-quaterphenyl the nematic phase is only metastable, *p-quinquephenyl* and *p-sexiphenyl*, and probably also *p-heptiphenyl*, have stable nematic phases [315]. Beyond p-quinquephenyl smectic phases are expected and proven for p-sexiphenyl. The magnitudes of the transition entropies correspond to the general observations on liquid crystals [1].

The series of p-oligophenyls illustrates thus, as shown in Fig. 5.1, that transition from the rigid crystal to the condis crystal can go with a displacive or with an order-disorder transition. Larger motion can occur with little correlation and thus

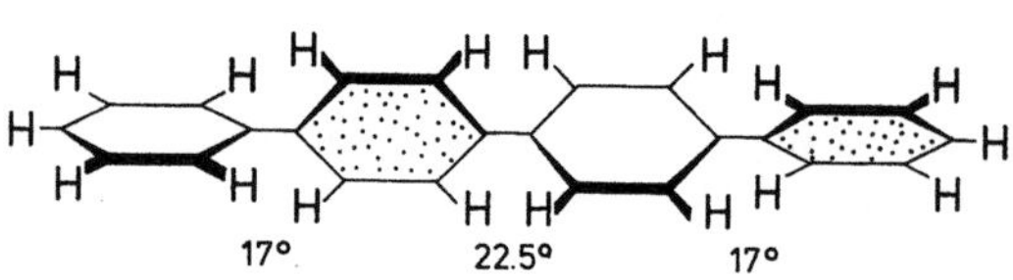

Fig. 5.4. Schematic drawing of the low-temperature conformation of quaterphenyl

gradual to such a degree that no thermal transition is noticeable. Changing features of the molecule, crystal and motion in the homologous series seem to make the various transitions more or less correlated. Transitions to the liquid crystals are, in contrast, always clear first order transitions with large entropy changes.

5.1.4 Polyparaphenylene

Polyparaphenylene is a rather intractible polymer with most of its thermal transitions seemingly above the decomposition temperature. Heat capacity measurements [317] are available up to 300 K only, and show complete agreement with the computed heat capacity based exclusively on the known vibration spectrum [318]. Unfortunately there are, however, no more direct measurements of the motion in polyparaphenylene available as yet. The crystal structure, determined by electron diffraction at room temperature has shown similarities to the low-temperature crystal-form of the quaterphenyl [319]. The chains are not overall planar, but the phenylene groups are twisted out of the average backbone plane. This twist is not regular, but random [319]. Since there is thus evidence that there is no averaging of the phenylene rotations to the planar conformation, one might speculate that below room temperature the polymer is frozen to a glass (L.C. glass, see Fig. 1.1). It would be of interest to follow the motion in polyparaphenylene with NMR, X-ray or neutron scattering as well as heat capacity to higher temperatures to discuss a possible glass transition into the liquid crystal phase and to check if alignment to a condis crystal is possible. The data in Fig. 5.1 are at too low a chain-length to extrapolate safely to high molecular masses.

5.2 Low Molecular Mass Liquid Crystals

Low molecular mass liquid crystals, especially in their nematic polymorphs, are characterized by close to liquid-like motion (see Sect. 1). Structurally they consist usually of the rod- or disk-like mesogen and one or more flexible alkyl or alkoxy groups. In the liquid crystalline state the conformational disorder and mobility is quite similar to that of the liquid state [11]. It remains thus to be explored if positional and conformational ordering occurs together in one step in going to the crystal, or in two stages in going first to a condis crystal (see Fig. 1.1). The condis crystal would then be of interest to the present discussion. Out of the large volume of low molecular mass liquid crystals [320] several examples have been selected to illustrate the major points of conformational disorder and mobility.

5.2.1 Molecules with Rod-Like Mesogens

Many molecules with rod-like mesogens have been analyzed in the liquid crystalline phase, but only few of the low-temperature precursor crystals have been studied. Among them one might find many conformationally disordered crystals [320]. A survey of such precursor crystals leads to three possible types of general disorder. Namely, packing defects, ring-flips or jumps, and disorder in the flexible ends. The latter two would lead to condis crystals of the types described for polyphenyls (Sect. 5.1) and paraffins (Sect. 4.1), respectively.

It was shown early that the rod-like molecules of liquid crystals are imbricated in their crystals [321]. For *PAA* and *PAP** is was observed that the positioning along the molecular chains may be variable, leading to streaking in the corresponding X-ray diffraction spots [321]. These packing defects seem to be strongly dependent on crystallization conditions and relate to the ultimate ease of mobility in the nematic phase. Molecules that are sufficiently symmetric show commonly disorder with respect to directional placement. Typical examples are the random directional placement of —N=CH— or —NO=N— in benzylideneaniline and azoxy compounds, respectively.

The second type of disorder refers to the phenylene groups and their bridging atom-groups, the twist of the phenylene groups adjacent to such central groups is governed by a balance between conjugation, steric hindrance and crystal packing. Quite analogous to the large-angle librations in polyparaphenyls (see Sect. 5.1) does one find sometimes planar X-ray structures for molecules that should have their energy minimum in a non-planar conformation. Condis crystals that interchange conformations by jumping between two rotational minima as indicated in Fig. 5.3 are then possible. For example, *CIBCIA* should according to molecular conformation and spectroscopic arguments have a 45° twist about the =N-φ- and an opposite 10° twist about the =CH-φ bonds. In its metastable, triclinic ($P\bar{1}$) crystal form, X-ray diffraction leads, however, to a planar conformation [322]. This planar conformation must have a 6–8 kJ higher intramolecular energy, but, as in the polyphenylenes, a high (±9° RMS) libration was reported and this suggests that there may be a low-temperature crystal-form with non-planar molecular structure.

A third type of disorder concerns molecules with longer, flexible end-chains. In these molecules one has the possibility of conformational disorder as described for paraffins and polyethylene (see Sect. 4.1). As an example, the series of *p-n-alkoxybenzoic acids* (*OBA*, see Table 5.2) may serve. In the crystals at room temperature the molecules form centrosymmetric, hydrogen-bonded dimers with all-trans CH_2-groups, except for a gauche-bond between the first and second carbon beyond the ether oxygen. The latter imparts a non-linear, non-planar geometry to the molecule. In the crystal the dimers form imbricated, stratified layers in which the aromatic nuclei and the alkyl chains are segregated into separate, close-packed strata. Antiparallel interdigitation of the alkyl chains of adjacent layers gives the 2:1 ratio of the chains to phenylenes in the strata, needed to achieve close packing [323]. From the p-hexyl to the p-decyl OBA two or more crystal polymorphs have been reported [323,324]. Table 5.3 shows the transition temperatures and entropies for two of these compounds as derived from differential scanning calorimetry [325]. Quite clearly the largest disordering occurs on going from the low-temperature crystal to the high-temperature crystal, the latter being most likely a condis crystal. The measured disordering entropies are more than 50% of the entropies of fusion of octane and decane [95.4 and 118.2 J/(K mol)]. All further transition entropies combined are somewhat less than commonly found for the fusion of a single, rigid, non-spherical motif (see Sect. 1). Such overall entropy of fusion of the condis crystal would be expected since the dimeric nature of the molecules should continue to some degree in the molten state.

* Full structures, names and transition temperatures are given in Table 5.2.

Table 5.2. Structural formulae and transition temperatures of some liquid crystals

PAA	Paraazoxyanisol T_d=390 K T_i=409 K 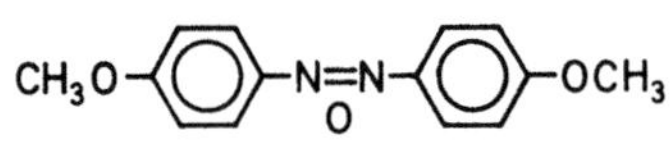
PAP	Paraazoxyphenetol T_d=409 K T_i=440 K 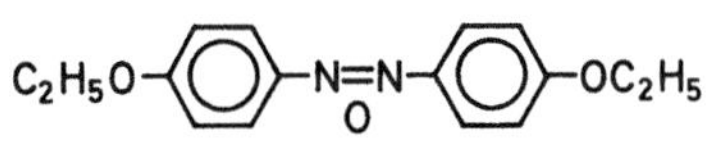
ClBClA	N-(4-chlorobenzylidene)-4-chloroaniline T_d none T_i=388 K 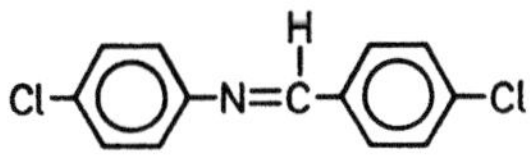
OBA	Para-n-alkyloxybenzoic acid (dimer) T_d and T_i see Table 5-3 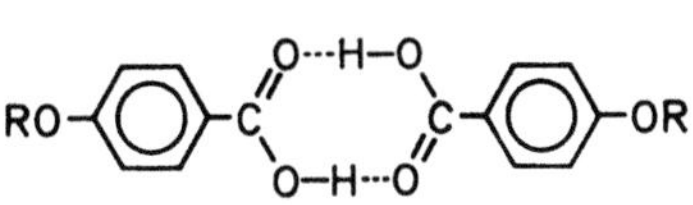
OOBPD	N,N'-bis (4-n-octyloxybenzal)-1,4-phenylene diamine T_d and T_i see Table 5-4 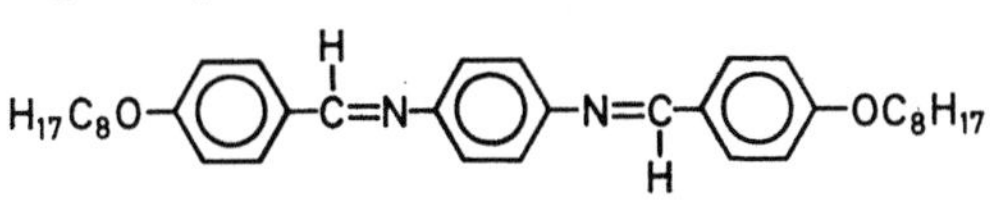
MBBA	N-(4-methoxybenzylidene)-4-n-butyl aniline (R=CH_3-) T_d=294 K T_i=320 K
EBBA	N-(4-ethoxybenzylidene)-4-n-butyl aniline (R=C_2H_5-) T_d=306 K T_i=349 K 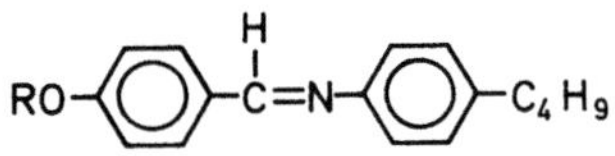
OHMBBA	N-(2-hydroxy-4-methoxybenzylidene)-4-n-butyl aniline T_d=317 K T_i=337 K 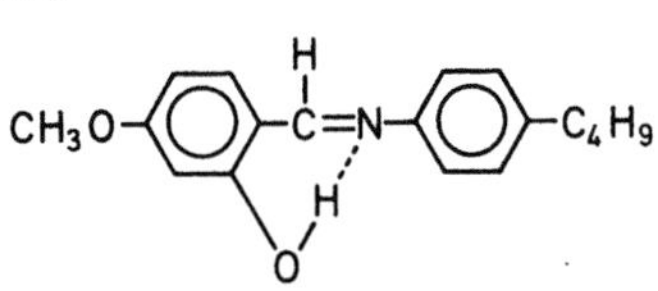

Table 5.3. Transition temperatures and entropies of transition for two p-n-alkoxybenzoic acids

	T_d (K)[a]	ΔS_d [J/(K mol)]	T_i (K)	ΔS_i [J/(K mol)]
p-n-octyloxy	348 (c)	51.7		
	374 (s)	29.1		
	381 (n)	3.3	419	6.0
p-n-decyloxy	359 (c)	60.6		
	370 (s)	28.83		
	398 (m)	4.2	416	7.0

[a] The first disordering goes to the condis phase (c), the second to the smectic phase (s), the third to the nematic phase (n). Data from Ref. [325]

Another example of the existence of condis crystalline polymorphs between liquid crystal and crystal phase is *OOBPD*. Table 5.4 contains a listing of the observed transitions [11]. Again, the combined entropies of transition from the condis crystal K1 to the melt [41.6 J/(K mol)] are what is expected for fusion of a single, rigid, non-spherical motif (see Sect. 1).

More direct studies of motion in this group of molecules by various types of NMR were often limited to the liquid crystalline state [32]. A largely motionally averaged spectrum for rotation about the molecular director and conformational disorder including in many cases rotation of the phenylenes about their para-axis is observed

Table 5.4. Transition temperatures and entropies of transition for OOBPD

Phase[a]	T_d (K)[b]	ΔS_d [J/(K mol)][b]
K3		
	385	27.3
K2		
	388	18.3
K1		
	415	16.1
H′		
	422	5.0
G′		
	427	0.9
I		
	437	7.4
C		
	475	7.6
N		
	505	4.6
Melt		

[a] Common notation of the phases: K1 to K3 are condis crystals, H′ to C smectic and N nematic liquid crystalline, see Ref. [11];

[b] For the N to Melt transition substitute T_i and ΔS_i, respectively

for the nematic phases [326–328]. The spectrum broadens then on crystallization due to the anisotropy of chemical shielding in the crystal.

More specifically, motion in the glass, crystal and nematic phase of *PAA*, *MBBA and OHMBBA* were studied over the full temperature range [329]. The calculated second moment of the proton NMR spectrum of OHMBBA is 19.5 G^2 for the rigid state, 15 G^2 when rotating CH_3— endgroups are present, 11.7 G^2 for additionally reorienting CH_3—O—, CH_2— and CH_3—CH_2— groups, and 3.3 G^2 for a molecule that has only its molecular axis oriented. The actually measured second moments did not reach 19.5 G^2 on cooling to 100 K, the lowest temperature measured. Both glass and crystal of OHMBBA and also MBBA showed at that temperature second moments of about 12 G^2, indicative of reorientation of more than just the last atoms in the end-chains. A gradual decrease in second moment is then observed in the solid state, to reach 4 G^2 for the glass at the glass transition temperature (208 K) and 7 G^2 for the crystal at the disordering transition (317 K). Also the T_1-relaxation times of glass and crystal were observed to be similar in their temperature dependence. The activation energies linked to end-group reorientation were measured to be 5.8 kJ/mol for MBBA and OHMBBA glass, 12.2 kJ/mol for MBBA crystals and 8.4 kJ/mol for OHMBBA crystals. As the glass transition and crystallization is approached from the nematic phase, the satellite peaks due to the motionally averaged phenylene hydrogens disappear (broaden), indicating slowing down of the phenylene jumps and flips. This rather high mobility as measured by NMR has led to the suggestion that the crystals of MBBA may actually be plastic crystals [330]. The entropy of disordering for MBBA [61 J/(K mol) [331]] is, however, much too high to permit such assignment. The total entropy of fusion of a plastic crystals should be only 7–14 J/(K mol) (see Sect. 1). The MBBA crystal, as well as EBBA and OHMBBA, are more likely condis crystals. Two metastable phases of MBBA with even higher mobility and indications of a glass transition have also been reported [329, 331–333].

5.2.2 Molecules with Disk-Like Mesogens

The two disk-like molecule types to be discussed are drawn in Fig. 5.5. The central benzene or triphenylene group makes up the rigid, disk-like mesogen and the six alkanoate groups represent potentially flexible chains containing in them actually more atoms than either mesogen. Sorai, Suga and coworkers [334–337] have made a detailed calorimetric and infrared study of four such compounds. They found that characteristically these molecules display at increasing temperature polymorphs of increasingly disordered nature in the paraffin ends. Conformational melting of the paraffin ends proceeds successively, to reach practically completion before reaching the transition to the liquid crystal or melt phase. These polymorphs are best classified as condis crystals.

Benzene-hexa-n-hexanoate (n = 5 in Fig. 5.5) has no discotic liquid crystal phase, but three condis states (phases III, II, I) [334]. The thermodynamic transition parameters are listed in Table 5.5. According to Eq. (1), one expects for fusion of the rigid crystal (phase IV) besides the ΔS_{pos} and ΔS_{orient} 24 conformational contributions [assuming the ester group O—CO— is rigid; about 228 J/(K mol)]. Indeed, the overall entropy of fusion is 286.4 J/(K mol), of which almost all conformational contributions [195.5 J/(K mol)] are already absorbed on heating to 348.3 K (phase I). In accordance with this simple analysis, IR data reveal that phase I is in it paraffinic components

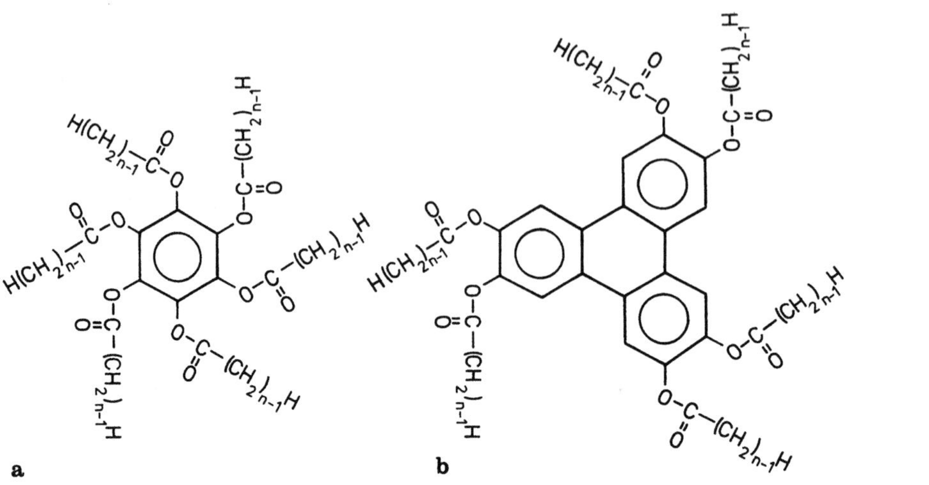

Fig. 5.5a and b. Chemical structure of discotic benzene and terphenylene hexa-n-alkanoates (**a** and **b**, respectively)

as disordered and mobile as the melt. On cooling, successive bonds are crystallized yielding condis crystals with less conformational disorder in phases II and III [334]. On reaching phase IV, the last two bonds take on fixed conformations.

The *heptanoate and octanoate* show liquid crystalline discotic phases over the narrow temperature ranges, 354 to 360 and 355 to 357 K, respectively [335,336]. The main observation is that the sum of the transition entropies in these two compounds is far less than expected from an analysis as just carried out for the hexanoate.

Table 5.5. Transition temperatures and entropies of transition for discotic molecules

Compound	T_d (K)	ΔS_d [J/(K mol)]	T_i (K)	ΔS_i [J/(K mol)]
Benzene-hexa-n-hexanoate				
	251.6	102.7		
	291.5	46.1		
	348.3	46.7	368.7	90.9
		195.5		
Benzene-hexa-n-heptanoate				
	129			
	222.8	59 (for 129, 222.8 and 230.8 together)		
	230.8			
	353.8	91.1	359.3	59.9
Benzene-hexa-n-octanoate				
	301.9	164.0		
	355.1	129.8	357.0	53.8
Triphenylene-hexa-n-octanoate				
	362.6	66.8	402.2	9.0

A comparison of the absolute entropies in the liquid phase shows that the progression from homologue to homologue is, however, what one expects for the addition of six CH_2-groups. The liquid crystal-forming compounds have thus a higher increase in entropy between zero and 250 K than expected for the additional vibrational entropy of the extra CH_2-groups alone. One must speculate that conformational disorder and moblity is introduced for these molecules through non-cooperative, local motion, i.e. the rigid crystals change gradually into condis crystals, beginning at temperatures below the transitions (see Sect. 2).

The *2,3,6,7,10,11-hex-n-octyl triphenylene*, finally, shows the largest deficit in conformational transition entropy collected in sharp, first order phase transitions [337]. In addition, this compound shows several polymorphs (at least 3) with largely different Gibbs-energy slopes that can easily be brought into metastable temperature ranges. For polymorphs III and I the entropy difference reaches 53 J/(K mol) at the cross-over point (290 K). Only data for the stable phase at high temperature (polymorph I) are listed in Table 5.5. Again, much of the conformational disorder that was discovered by infrared analysis must be acquired through non-cooperative, local motion at low temperature. Depending on the packing in the crystal, which is determined by the crystallization history, different polymorphs, and thus different rates of acquisition of conformation entropy are observed. Much more strucutral and motional information is needed to unravel the details.*

5.2.3 Amphiphilic Molecules

Amphiphilic molecules consist of at least two sufficiently large parts that would, if not connected by covalent bonds, separate in the liquid state into different phases (*amphi* — Grk. of two kinds, and *philos* — Grk. loving). Most well-known are amphiphilic molecules of polar and non-polar parts such as, for example, long-chain aliphatic hydrocarbon segments connected to an acid, ester or salt. Simular conditions can also be produced between two or more sufficiently long, incompatible non-polar parts such as aliphatic hydrocarbons, aromatic hydrocarbons, perfluorohydrocarbons or siloxanes. In the liquid state such molecules will organize with well-defined interfaces in which the junctions between the incompatible parts are collected. Depending on concentration, lamellar, rod-like or spherical supermolecular structures are common for such molecules. It is possible to add a solvent, compatible with only one of the molecular segments and separate the superstructures into micelles. Typical examples of such mesophases are soaps, lipids and block-copolymers (lyotropic liquid crystals) [320]. The function of the rigid mesogen to produce partial orientational order is in these cases taken over by the interface. For the present discussion it is of interest to explore if besides the liquid crystalline states there exist on cooling recognizable, stable condis crystalline states before complete solidification to the glass or crystal (see Fig. 1.1).

As one might expect from the discussion of the liquid crystals in Sects. 5.2.1 and 5.2.2, condis crystals may result whenever a paraffin chain is attached to a part of

* Looking just at the entropy data of Table 5.5 one could also suggest that the gradually produced condis crystal [ΔS_{conf} = 342 J/(K mol)] changes at 362 K into a plastic crystal (ΔS_{orient}) and melts at 402 K (ΔS_{pos}). Perhaps proper placement of the alkanoate groups about the triphenylene might lead to a more spherical than disk-like mesogen?

Table 5.6. Transition temperatures and entropies for n-alkyltrimethylammonium halides[a] n-$C_nH_{2+1}N^+(CH_3)_3X^-$

Compound	T_d (K)	ΔS_d [J/(K mol)]
n = 10 X = Br	369.5	88
n = 12 X = Br	370.4	114
n = 14 X = Br	372.3	126
n = 16 X = Br	376.1	138
n = 18 X = Br	378.3	172
n = 22 X = Br	382.6	218
n = 12 X = Cl	357.2	83
n = 14 X = Cl	364.5	103
n = 16 X = Cl	368.4	113
n = 18 X = Cl	374.7	114
n = 18 X = I	393.9	128

[a] Data from Ref. [340], T_i at about 530 K under decomposition

the molecule that retains long-range, crystalline order to a higher temperature than the transition to the hexagonal paraffin phase.

One of the first observations of such stepwise, separate phase transitions of the paraffinic part of an amphiphilic molecule was made on *tetra-n-amyl ammonium thiocyanate* [338]. At 315 K the crystals undergo a major first-order transition [ΔS_d = 71 J/(K mol) with a 4.8% volume change]. The final transition to the isotropic, molten salt at 323 K is entropically smaller [ΔS_i = 59 J/(K mol) with a negligible volume change]. Since typical alkali thiocyanate entropies of fusion are 30–40 J/(K mol) [339] it is reasonable to assume that the high-temperature crystal state is that of a condis crystals with considerable conformational freedom.

A series of *n-alkyltrimethylammonium halides* that shows a well-resolved condis state has been described recently [340]. The results of thermal analyses on these compounds are listed in Table 5.6. The entropy of disordering changes by about (n—1) times the typical conformational entropy of fusion per CH_2-group (see Sects. 1 and 4.1). X-ray diffraction on these compounds revealed the complete loss of the hydrocarbon lattice at T_d, but a remaining ionic crystal structure. Higher temperature transitions could not be observed on these compounds because of the beginning of decomposition.

The simpler *n-alkylammonium chlorides* show a similar behavior [341,342], just that the degree of disordering within the transition is much less. Table 5.7 displays the results. The phase that exists below the temperature of change to the isotropic liquid was identified as a smectic phase, soap (see below). Then entropy change on going to the isotropic phase is typically low, as usually seen in liquid crystals. The two disordering transitions listed [crystal (?) to condis phase and condis phase to smectic phase] do not add-up to complete conformational disordering (compare with Table 5.6). Either there are further low temperature transitions, as was found for n = 15 [341], or conformational freedom was gained without cooperative transition, as seen for discotic liquid crystals (see Sect. 5.2.2). It is thus not certain that the phase below the first T_d listed is a crystal. It may still show dynamic, conformational disorder. X-ray analysis showed typical bilayer structures for all phases. Even after

Table 5.7. Transition temperatures and entropies for n-alkylammonium chlorides[a] n-$C_nH_{2n+1}NH_3^+Cl^-$

Compound	T_d(x)	ΔS_d [J/(K mol)]	T_i (K)	ΔS_i [J/(K mol)]
n = 9	325	9.7		
	471	17.1	542	3.3
n = 10	321	13.0		
	466	17.6	533	3.1
n = 11	339	8.5		
	461	19.2	521	2.8
n = 12	347	8.4		
	458	20.0	512	2.4
n = 13	359	11.1		
	456	22.6	503	2.5
n = 14	365	11.8		
	450	23.4	493	2.2
n = 15	374	15.2		
	447	25.0	486	2.0
n = 18	386	20.4		
	436	23.8	466	1.7

[a] Data of Ref. [342]; DSC curves were only registered from 270 to 600 K and lower temperature transitions may have been omitted; for example for n = 15 in Ref. [341] one can find additional solid-solid transitions at 326 and 346 K with entropy changes of 9.8 and 17.1 J/(K mol)

going in temperature above the isotropic phase (clearing point) the layered organization persists [342]. As judged from the increase in the X-ray long-spacing on going from n = 10 to n = 15, one finds a $\Delta\bar{c}$ per two CH_2-groups of 0.19 nm for the condis crystal, and 0.13 nm for the smectic crystal. These values compare with 0.25 nm for polyethylene in the all-trans conformation [47]. The lateral distances do not increase sufficiently to let the hydrocarbon chains approach a random conformation, i.e. the naming of the disordered mesophase a "plastic" phase [341] cannot be correct. There is no indication of a close to spherical mesogen. In addition, the entropy for transition into the smectic phase of 17–25 J/(K mol) is somewhat high for a plastic-to-liquid crystal transition. Finally, it would be unreasonable after losing orientational order in a plastic phase to gain it again in heating to a smectic phase. Classification of the disordered phase as a condis phase seems thus more correct.

The thermal behavior of *thallium(I) salts of aliphatic carboxylic acids* of two to 22 carbon atoms were analyzed [343] and classified according to their miscibility [344]. The first member of the series, CH_3COOTl, shows a one-step melting transition to the liquid phase [T_m = 402.2 K, ΔS_m = 43.5 J/(K mol)]. Already the three and four carbon homologues reveal a strong decrease in ΔS_m to 18.8 and 15.8 J/(K mol) at 462.4 and 451.9 K, respectively. The non-polar part of these short molecules is still too small to show any liquid crystalline phase, but their crystal structure must be much more disordered than that of thallium acetate. All other thallium carboxylates show as high-temperature crystal-phase a liquid crystalline "neat" phase with fan texture under polarized light [ΔS_i = 7 — 2.4 J/(K mol) with increasing number of carbon atoms]. The entropy of transition into the neat phase changes from about 10 J/(K mol) to 60 J/(K mol) on going from five to 22 carbon atoms, still much too little to assume full order in the lower-temperature crystal-phases. Indeed, for all carboxylates with more than 9 carbon atoms, two and possibly three condis phases

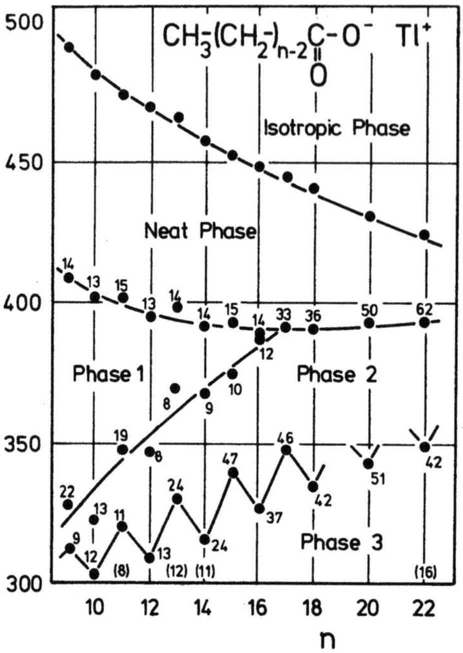

Fig. 5.6. Transition temperatures of thallium carboxylates. Numbers at the various points indicate transition entropies in J/(K mol). Numbers in parenthesis indicate a further low temperature phase of the given entropoy of transition. Drawn after Ref. [344]

are indicated (Phases 1,2) that show additional disordering entropies of similar magnitude to three times the entropy change into the neat phase. Figure 5.6 illustrates the phase relationships.

Carboxylic acid salts of alkali and alkaline earth metals are called *soaps*. Their structure is well analyzed, but most emphasis was placed in the past at their relation to the smectic phases and their structure in lyophilic liquid crystals [320]. A reexamination of *lithium n-hexadecanoate* [345] revealed, in analogy to the ammonium salts, that the stepwise fusion of the double-layer crystal structure may well go from crystal to condis crystal to smectic phases. Below 384 K lithium n-hexadecanoate shows a typical bilayer structure (LC 1) for which infrared absorption suggests monoclinic or orthorhombic chain packing of all-trans chains [345]. At 384 K a sharp transition with 29.8 J/(K mol) entropy change occurs to what may be a condis phase (LC 2). The paraffin chains are still largely trans, but show now a more symmetrical (hexagonal?) packing. At 482 K this condis phase changes with a much broader transition [ΔS_d = 35 J/K mol)] to the two-dimensional rectangular phase [346] in which ribbons (about 4 nm wide) of rectangularly ordered bilayers of ions permit lateral disordering of the hydrocarbon chains. The phase is now "waxy" and remains birefringent. The "clear-point" occurs sharply at 504 K with an entropy change of 36.7 J/(K mol). As in the n-alkyl ammonium chlorides, smectic ordering is still present above the clear-point. The total transition entropy [101.5 J/(K mol)] is higher than found for the n-alkyl ammonium chlorides, but still, perhaps somewhat low for complete disordering.

A detailed thermal and X-ray analysis exists for a series of anhydrous *Na soaps* [347]. From these measurements it can be derived that the soaps with alkyl-chains of 8, 16 and 18 carbon atoms undergo a crystal to condis crystal transition at about 390 K with entropies of 14 to 20 J/(K mol) (loss of paraffin structure). The longer chain soaps follow this with a second transition at 408 K of 24.7 and 29.1 J/(K mol) entropy change, quite similar to the n-alkylammonium chlorides. A last, larger entropy transition (subneat to neat) occurs at 515 to 527 K with a ΔS_d of 15 to 19 J/(K mol). The final transition to the isotropic phase takes place at T_i values of 637, 564 and 552 K and entropies ΔS_i of 2.0, 1.5 and 1.4 J/(K mol) for the C_8, C_{16} and C_{18} soaps, respectively. It may well be that all phases between 390 and 530 K are condis phases (subwaxy, waxy, superwaxy, subneat in the usual nomenclature). The neat phase is certainly on the basis of the transition entropy magnitude to be classified as a smectic liquid crystal. The subneat phase would need more detailed motional and defect analysis to distinguish between a more highly ordered smectic liquid crystal (that would naturally possess conformational disorder) and a condis crystal (that would have no dynamic disorder of the mesogen, i.e. the carboxylate end of the molecule).

The structurally somewhat more complicated *lipids* may show in the pure state quite similar sequences of phase transitions as the soaps. Their behavior may be illustrated by the transitions of 1,2-dimyristoyl-DL-phosphatidylethanolamine: $[CH_3—(CH_2—)_{12}CO—O—]CH_2[CH_3(CH_2—)_{12}CO—O—]CH—CH_2—[O—PO_2^{(-)}—$ $—O—(CH_2—)_2—NH_3^{(+)}]$.

Thermal analysis, infrared, X-ray and îH NMR data were collected [348,349] and interpreted as follows: At about 390 K a sharp endothermic transition with most of the overall entropy-change indicates the transition into the condis phase of the paraffin chains. There is a second, small transition at about 408 K that may involve only a minor readjustment of the remaining crystal. At 487 K, finally the clearing is observed in a capillary melting point experiment. Only a minor endotherm corresponds to this transition to the isotropic phase. The low temperature phase is identified as fully ordered through X-ray diffraction, and the infrared spectrum shows the typical paraffinic crystal-splitting of the absorption lines. NMR-linewidth studies reveal at 77 K a second moment of about 27 G^2 that gradually narrows, and starts at about 185 K to develop a weak, narrow (3G line-width) second component. The rigid crystal approximation needs a second moment of about 29 G^2. The slightly lower value at 77 K and its decrease is easily accounted for by the end group ($—CH_3$) rotation. The narrow component is an indication of larger motion without loss of order below the transition. At the 390 K transition, the infrared spectrum becomes liquid-like for the CH_2-groups. The bilayer-spacing decreases (X-ray data) and the second moment of the NMR spectrum drops to 0.1 G^2. Such second moment indicates considerable mobility within the time scale of the NMR experiment, but is still several orders of magnitude larger than for a true liquid (10^{-4} to 10^{-5} G^2). Diffusion and rotation of the whole molecule is thus still largely inhibited. The paraffin chains are, as in the soaps in a dynamically conformationally disordered, a condis crystalline state. The hydrophillic part of the molecule keeps the molecule in a state of positional and orientational order.

n-Paraffin molecules with hydrophilic groups on both ends separate also on crystallization into layers of different molecular character. Compared to condis crystals of pure paraffins these molecules cannot translate in the chain direction to accommodate conformational defects in the interface (see Sect. 4.1). An analysis of *16-hydroxy lithium hexadecanoate* $[CH_2OH(CH_2)_{14}COOLi]$ by calorimetry, X-ray diffraction, and infrared spectroscopy showed a condis phase transition at 386 K [58.8 J/(K mol)] and the transition to the isotropic melt at 504 K [82.8 J/(K mol)] [350]. The condis phase is still of bi-layer structure. The infrared spectrum shows in the low-temperature crystal phase the typical polyethylene-like crystal splitting of the rocking and bending absorptions due to two non-equivalent chains. On transition to the condis phase, correlations between (and within) the paraffinic chain are lost, although bonds connected to defects, as found in condis parafins, in the 1300–1400 cm^{-1} frequency region were not found. It may well be that larger trans-segments rotate in unison [350].

A deviation from the paraffin behavior was also found for *1,20-eicosanedioic acid* $COOH-(CH_2-)_{18}COOH$ [166]. This compound has no condis phase at atmospheric pressure (T_m = 398 K), but shows a hexagonal condis phase at elevated pressure (T_d = 474 K at 580 MPa and 488 K at 680 MPa). Paraffins of similar chain-length are still decreasing their condis phase-area with increasing pressure (see Sect. 4.1).

Finally, one must mention a series of amphiphilic molecules made of segments of perfluorinated hydrocarbon and normal hydrocarbons [351,352]. Most data are available for molecules of the type $F(CF_2-)_{12}(CH_2-)_nH$ with n varying from 0 to 20. Figure 5.7 illustrates the changes in entropy of the two highest temperature transitions. The highest temperature transition is the change into the isotropic state with transition temperatures changing from 334 to 364 and finally to 369 K on going from n = 0 to 14 and finally 20, respectively. Up to n = 14 there is practically little increase in ΔS_i above the value for n = 0 which, in turn, corresponds well to the extrapolated value for ΔS_i of $C_{12}F_{26}$ [$1.5 \times S_i(C_8F_{18})$ = 62.4 J/K mol)]. The only explanation to this result can be that the hydrocarbon part is already almost completely molten and does not contribute additionally to ΔS_i. Since the hydrocarbon parts are linked by covalent bond to the fluorocarbon portion, they cannot, however, have translational or rotational freedom in this restrained molten state, i.e. they must be in a condis state. The fluorocarbon portion in itself is known to also be conformationally disordered (see Sect. 4.2) and indications of their low temperature disordering transition have ben found by calorimetry at T_d = 192, 202, and 211 K for n = 8, 19, and 12, respectively [352]. The disordering entropies ΔS_d illustrated in Fig. 5.7 are observed at 313 to 362 K for n = 4 to 14, respectively. For n > 14 the two transitions occur then at the same temperature with a combined entropy. The slope of ΔS_d is 3–4 J/(K mole CH_2), this is the same magnitude as found for the transition of condis polyethylene

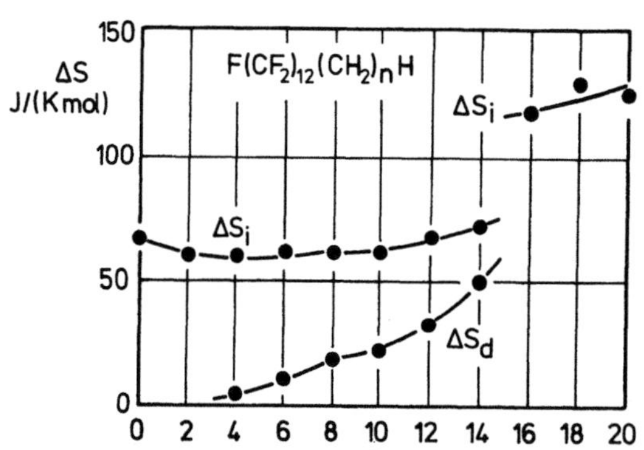

Fig. 5.7. Entropies of transition for partially fluorinated n-alkanes of the type $F(CF_2-)_{12}(CH_2-)_nH$ for values of n from 0 to 20. Drawn after Ref. [351]

to the isotropic phase (see Sect. 4.1). Raman spectroscopy and ^{13}C NMR (CP MAS) support this analysis. The Raman band-pattern below T_d is closely similar to the loose packing of hydrocarbon segments in the hexagonal phase [351]. The NMR spectrum of the inner CH_2-groups, in turn, resembles closely that of the condis phase of paraffins [352]. While the Raman spectrum does not change significantly on traversing T_d, there is a strong shift in the NMR spectrum that now closely resembles that of the liquid state. The overall extension of a large number of the molecules as measured by the frequency of the longitudinal accoustic mode of the Raman spectrum (LAM) does not change significantly until the isotropic state is reached. For the longer hydrocarbon sequences $n > 16$ there seems to be some evidence for a change to the orthorhombic packing below 270 K. Questions about the ultimate freezing of the conformational disorder are still open. It is not clear as yet if there is a continuous (local) introduction of the conformational order or a cooperative glass transition (see Fig. 1.1), or perhaps a further T_d.

5.3 Macromolecules with Rigid and Flexible Segments

Macromolecular liquid crystals can be obtained by sequencing rigid mesogens with flexible spacers. The latter provide the needed mobility for the mesogen to achieve partial dynamic orientational order [353]. It is possible to place the mesogen either in the main-chain or in the side-chain. In the liquid crystalline state residual order in the spacer diminishes with increasing distance from the mesogen [354]. As in the low molecular mass liquid crystals, conformational disorder is part of the liquid crystalline state (see Sect 1). To achieve a pure condis crystalline state, the translatory mobility and disorder of the liquid crystal must change to long range order, and the partial orientational order must improve, i.e. the mesogen and the flexible part of the molecule must obtain a fixed position in the crystal (positional and orientational order) while some dynamic conformational disorder must be maintained for the mesogen (see Sect. 5.1) and/or the flexible spacer (see Sects. 4 and 5.2).

The macromolecular nature complicates the study of the condis state due to the common limitation to small, and often defect, crystals [7,42,155]. The low degree of crystallinity of macromolecules makes it sometimes difficult to distinguish between long-range perfection in small crystals or the presence of larger volumes of less perfect mesophase crystals or glasses.

The distinction between macromolecular liquid crystals and condis crystals may be helped by experience, although the differences are usually only a matter of degree. Liquid crystals seem always to convert fully from the liquid state to the liquid crystalline state, and the conversion to the mesophase on cooling shows relatively little supercooling (few kelvins). Condis crystals, in contrast, show more often than not partial conversion from the melt and a relatively high supercooling (10 or more kelvins), i.e. they are much closer in their behavior to fully-ordered, macromolecular crystals. The crystallization kinetics of liquid crystals has found little attention in the past. It seems describable by Avrami expressions with exponents of 2 to 4 [1]. Condis crystals, in contrast, have in the few analyzed cases shown exponents of less than 2 [244]. While the transition entropy of the liquid crystal to the isotropic melt is always low and the change of heat capacity at the glass transition is similar to that of the liquid [1], the condis crystals transition entropy and also the change in heat capacity at the glass transition is variable. The actually measured transition parameters for the condis crystal do not only depend upon the degree of conformational disorder, but also on the degree of condis crystallinity.

In this section some examples of documented and suspected condis crystal phases in macromolecules consisting of rigid and flexible segments are summarized. Comparing the few examples with the abundance of condis crystals in analogous low mass molecules (Sect. 5.2) one can predict that there must be many more examples than are known at present. In Sect. 5.3.1 macromolecules that do not have sufficiently long rigid segments to exhibit a liquid crystalline phase, but show a condis phase are described. In Sects. 5.3.2 and 3 macromolecules with main-chain and side-chain mesogens are briefly reviewed as to their likelihood to have condis crystallinity at temperatures lower than the liquid crystalline phase area. Section 5.3.4, finally, contains a discussion of the condis states that result when flexible side-chains are attached to macroscopic mesogens. Macromolecules out of stiff chains are the substance for Sect. 6.

5.3.1 Polymers without Liquid Crystalline Phases

Polyparaxylylene (PPX) can be thought-of as a regular, alternating copolymer of segments of flexible CH_2—CH_2— and rigid C_6H_4—. Both parts of PPX give as pure homopolymers rise to condis crystals, but of largely different nature (see Sects. 4.1 and 5.1): Polyethylene, through dynamic introduction of gauche-conformations, and polyparaphenylene, through partial or complete ring-flips or rotation about the para-axis. Polyparaxylylene shows two disordering transitions before going into the isotropic state. Figure 5.8 illustrates the disordering transitions for a well annealed sample at 500 K [$\Delta S_d = 9.9$ J/(K mol) recalculated to 100% crystallinity] and 570 K [$\Delta S_d = 2.7$ J/(K mol), recalculated to 100% crystallinity]. The transition to the isotropic state occurs outside of the temperature limit of Fig. 5.8 at 700 K with a ΔS_i of 14.3 J/(K mol), followed by decomposition [61]. Almost 50% of the entropy of fusion is thus absorbed below the transition into the isotropic state. Based on the experience with polyparaphenylene and polyethylene it was suggested that the β_1-polymorph may show phenylene disorder [355]. On transition into the β_1-phase there is no change in the c-axis dimension and orienation [356]. The barrier height for rotation about the phenyl-CH_2-bond is known to be much lower than for the

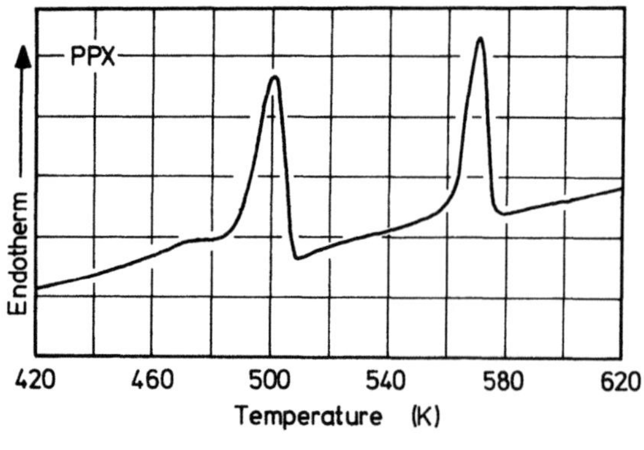

Fig. 5.8. Differential thermal analysis curve of a PPX sample polymerized from the cyclic dimer at 296 K and annealed below the transition to the isotropic state, at 625 K, for 12 h. The shoulder at about 475 K is an annealing peak (fusion of imperfect crystals). Transition α–β_1 occurs at 500 K (T_{d_1}). Transition β_1–β_2, at 570 K (T_{d_2}). The endotherm for T_i (700 K) is not shown (see Ref. [355])

CH_2—CH_2-bond, making phenylene mobility and disorder likely. The crystal structure of the fully ordered α-polymorph is monoclinic (C 2/m) [357] with parallel packing of the phenylenes, while the β_1-structure is known to have statistical defects in a much enlarged trigonal unit cell (P_3) [358], but the detailed nature of the defects is still in dispute [355,356]. The β_2-polymorph, ultimately must, in addition have CH_2—CH_2—disorder [355]. The hexagonal unit cell contains now only one chain and has, due to conformational motion and disorder, cylindrical symmetry, originally classified as a "rotor phase" [358]. Much remains to be done to resolve the motion and disorder in the high-temperature phases in polyparaxylylene. The research is particularly hampered because of some rather surprising structure changes discovered through changes in the transition peaks on thermal treatment. Incomplete, delayed, connected and early transitions were observed involving the two peaks shown in Fig. 5.8 [355]. It seems rather easy to freeze-in different degrees of disorder (CD glass transitions, see Fig. 1.1) despite the rather low glass transition temperature of the amorphous polymer (286 K).

Similarly confusing thermal analyses with strong indications of condis phases instead of, or in addition to, the reported liquid-crystalline phase were found for *poly(7-oxabicyclo[2.2.1]heptane-co-tetrahydrofurane)* $(C_6H_{10}—O—)_x[(CH_2—)_3O—]_y$ [359]. Multiple, ill-defined endothermic transitions between 370 and 700 K are reported for fibers with seemingly little correlation to the spinning properties. The two components of the copolymer can undergo isomorphous (or isodimorphous) substitution in the crystals so that for all ratios of x to y crystallinity remains possible. The expected conformational disorder resides in the aliphatic sequences in analogy to polyethylene (see Sect. 4.1) and in the cyclohexane ring as described in Sect. 3.1.1. Since the ultimate transition to the isotropic phase occurs only at temperatures of up to 650 K, both components are in the temperature range were motion is possible and conformation disorder should be favored by the copolymer disorder. Inspecting the published X-ray diffraction patterns [359] shows a largely concentration-independent, complete disappearance of all 0.3 to 0.4 nm diffractions at 370 K, parallel with a major endotherm. This may be evidence for conformational disorder in the tetrahydrofurane sequences. Again, much more detailed motion and structure information is needed to understand this interesting polymer.

Finally, one may want to look at *poly(ethylene terephthalate)* and the homologous longer CH_2-sequence polymers. Here, again, a relatively large, rigid group (CO—C_6H_4—CO—) is in sequence with flexible chain segments [O—$(CH_2—)_x$O—]. Under elevated pressure crystallization to an extended chain crystal morphology has been observed [360]. Although no special high temperature and pressure polymorph of high degree of disorder and mobility has been reported, one must assume that under these conditions higher mobility, and possibly also conformational disorder, exists. Proton and ^{13}C NMR (MAS) between 100 and 500 K showed that at atmospheric pressure a rigid crystal exists only up to about 100 K [361]. Angular fluctuations of both the CH_2-sequences and the phenylene groups in amorphous, crystalline and rigid amorphous portions of the polymer reach increasingly larger amplitudes and frequencies (10^9 Hz, 20° at 430 K, for the crystalline areas, for example). In addition, it was suggested that trans-gauche conformational interchanges and 180° C_6H_4-ring-flips may be possible [361]. All this could lead to the needed high mobility for extended chain crystallization. Even larger amplitudes of motion in the CH_2-sequences were found in poly(butylene terephthalate) [362], involving a three-bond motion of the carbon sequence (gtg). With relatively fixed phenylene groups, the two central CH_2-groups can undergo relatively large amplitude excursions.

5.3.2 Liquid Crystalline Polymers with Main-Chain Mesogens

Liquid crystal forming polymers with flexible spacers in the main chain have found major attention in the last years because of their non-linear optical properties and potential ease of orientation to high modulus fibers. The structure principle, synthesis and properties of such macromolecules have been reviewed recently [363]. Of special interest to the discussion of the condis state is the possible conformational disorder and mobility of such molecules at temperatures below the disordering transition, T_d into the liquid crystalline state. Figure 5.9 illustrates the typical thermal behavior of such a polymer, quenched into the LC glass state [364]. In principle, the curve is similar to the analogous low molecular mass molecules described in Sect. 5.1.1 (see also Table 5.2). Quantitatively the heat of transition from the crystal to the liquid crystalline phase is, even after extrapolation to complete crystallization, much lower than expected for a fully ordered crystal [1]. The example of Fig. 5.9 reaches for 100% crystallinity, a ΔH_d of only 13.5 kJ/mol. This contrasts about 65 kJ/mol estimated from the collective experience on heats of fusion of normal polyesters and 31 kJ/mol measured for p-azoxyanisole (see Table 5.2). Based on such findings that are duplicated qualitatively on over 50 other main-chain liquid-crystalline polymers in similar magnitudes [1], one must conclude that there is the possibility of considerable conformational disorder below T_d. As for low molecular mass liquid crystals described in Sect. 5.2 there are sometimes distinct, low-temperature transitions of high entropy, indicating that a further ordering of the condis crystals may occur in stages [365].

An interesting observation in the main-chain liquid crystals was an enhanced

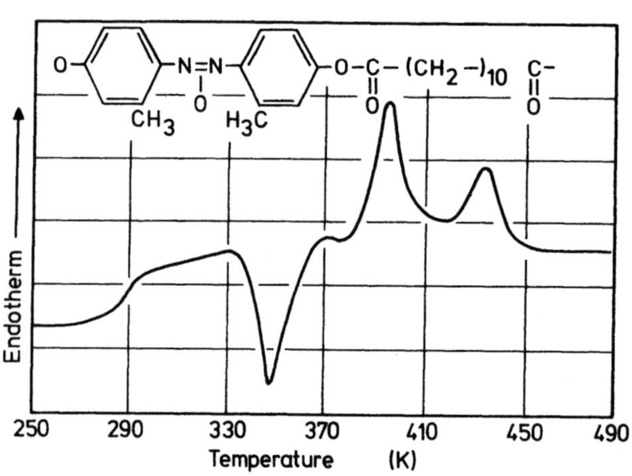

Fig. 5.9. Differential thermal analysis curve of a liquid crystalline polymer with flexible spacers in the main chain [Poly(oxy-2,2′-dimethylazoxybenzene-4,4′-diyloxydodecanedioyl)]. Heating at 50 K/min after cooling at 50 K/min. LC glass transition at 290 K. Crystallization of the mesophase at T_c = 345 K. Disordering transition at T_d = 385 K. Transition to the isotropic phase T_i = 420 K, ΔS_i = 9.8 J/(K mol). Drawn after Ref. [364]

odd-even effect of the flexible spacer length on the disordering temperature and transition temperature to the isotropic state [366]. Deuterium magnetic resonance enabled the quantitative evaluation of conformational disorder and mobility responsible for the low heats of fusion and the higher level of disorder in the odd spacer-atom molecules [367].

Figure 5.10 illustrates the change in fraction of trans-conformations in the liquid crystalline and condis crystalline states of two polyesters, as determined by multi-pulse, dynamic deuterium NMR. For this model-system a more complete analysis is available [368-370] and it may serve to illustrate the major points of mobility and disorder. The uniform 20% gauche conformations in the crystal phase indicates a likely high degree of disorder since typical crystals with long sequences of methylene groups have, at best, isolated gauche conformers at the ends [47]. Thermal analysis places the glass transition at 303 K and T_d occurs at 429 and 433 K and T_i at 543 and 553 K for n = 9 and 10, respectively [369]. The measured transition enthalpy ΔH_d is even lower than ΔH_i (about 4.5 and 7.0 kJ/mole, respectively) [370] a fact that is in accord with the high gauche population. NMR line shape and relaxation analyses between 130 and 393 K reveal clearly separable liquid crystalline (mobile) and condis crystalline (immobile) fractions (55 ± 5% for the latter). Only the liquid crystalline fraction shows diffusive motion of larger chain segments. Both fractions show local conformational trans-gauche isomerizations, but with different dynamic ranges. In the condis phase correlation times range from 1.5 to $5 \cdot 10^{-6}$ s for the isomerization and in the liquid crystalline phase, from $5 \cdot 10^{-9}$ to $8 \cdot 10^{-8}$ s on changing in temperature from 393 to 255 K. The activation energy for the trans-gauche isomerization is 6.7 kJ/mol for the condis crystals.

With defect concentration and dynamics in the flexible spacer known in the model system of Fig. 5.10, order parameters are of interest. Deuterium NMR could in this case establish the micro-order parameter S_{zz} of the molecular segment z-axis with

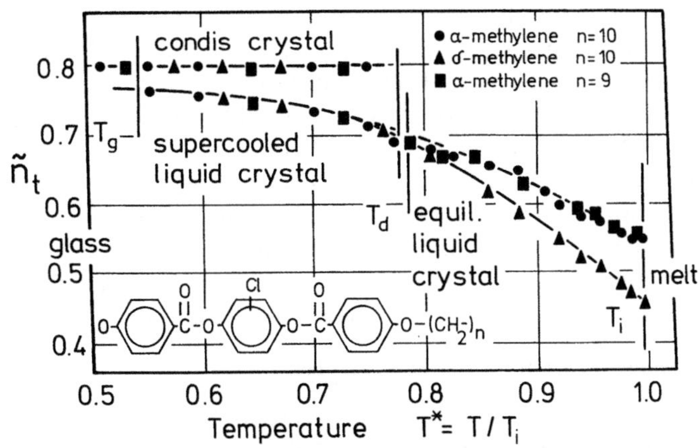

Fig. 5.10. Temperature dependence of the fraction of gauche conformations in the alkyl portion of a macromolecule as determined from suitable deuterated samples by multiple pulse dynamic deuterium NMR. Data from Ref. [368]. Poly[1,4-oxybenzoyloxy-1,4-(chlorophenoxy)carboxy-1,4-phenoxy(deca or nona) methylene]. The transition to the isotropic phase at T_i = 543 K for n = 9 and 553 K for n = 10, ΔS_i about 13 J/(K mol)

respect to the director axis [369]. For n = 10 (see Fig. 5.10) the condis crystals reach the high order-value of $S_{zz} = 0.9$ and for n = 9, a much lower value $S_{zz} = 0.54$ was found. In the liquid crystalline phase the odd-even differences are much less. At T_i, the order parameter S_{zz} for molecules of n = 10 jumps from zero to 0.65, to reach 0.87 at T_d to then level-off. For molecules of n = 9 the respective values for S_{zz} are 0.55 and 0.80.

The order in the liquid crystalline phase of the macromolecules with main chain mesogen is much higher than in analogous, low molecular mass liquid crystals. These special properties suggest a possible path to high-strength fibers since the orientation of the director axis in the magnetic field and also on mechanical drawing can be extremely high [370]. It is of interest that the long known hard-soft segmented polymers give quite similar data on NMR analysis [371]. They have a similar, but usually less regular structure that could also be used as a starting point for high strength, high orientation products. The hard segments, usually constructed of aromatic polyurethanes or polyesters, are glassy or crystalline phases at use-temperature and the soft segments, usually constructed of aliphatic polyoxides or polyesters, are ranging from rubbery to condis crystalline at use temperature. A liquid crystalline phase at elevated temperature could allow more perfect orientation.

Another way to reach flexibility of polymers with main-chain mesogens is the use of not quite rigid mesogens. This method was recently developed by employing cyclosiloxanes as mesogens [372–374]. As monomers the cyclosiloxanes with eight or more chain atoms show large conformational mobility down to low temperatures (< 140 K). Only when restricted by a crystalline or glassy environment is this motion stopped (see Sect. 3.3). Two polycyclosiloxanes produced had the following structure with n = 1 (A) and 2 (B):

$$\left[-(CH_3)Si\begin{matrix}\diagup O-[Si(CH_3)_2-O-]_n\diagdown \\ \diagdown O-[Si(CH_3)_2-O-]_n\diagup\end{matrix}Si(CH_3)-O-\right]_x$$

another polymer with n = 2 had the bridging Si's substituted with phenyl groups (C). All three types have low temperature glass transitions, T_g = 222, 182, and 232 K for A, B, C, respectively, independent of isomerism of bridge substitution (cis or trans). The transition into the isotropic state varies with isomeric purity. The reported temperatures T_i are 355 K for n = 1 (A), 100% trans; 525 to 595 K for n = 2 (B), 75% cis to 85% trans; 470 to 700 K for n = 2 (C), 98.5% cis to 99% trans. Measured values for ΔS_i are 1.4 to 8.1 J/(K mol), typical for liquid crystals. Compounds with high isomeric purity show a disordering transition, compounds with low isomeric purity show only the glass transition. The interplay of conformation, ring-flexibility, crystallization, and glass transition has not been studied in detail as yet.

5.3.3 Liquid Crystalline Polymers with Side-Chain Mesogens

Liquid crystal forming polymers with flexible spacers in the side-chain were developed as a logical consequence of the analysis of side-chain crystallization. With sufficiently long side-chains, starting with about 10 chain atoms, the crystallization behavior becomes increasingly similar to low molecular mass molecules, i.e. the macromolecule backbone becomes only a connecting backbone with little consequence for the thermodynamics and kinetics of crystallization [7,375]. Detailed reviews of synthesis, structure and properties of liquid crystalline macromolecules with mesogens in the side-chain were published recently [376–379]. For the present purpose it is only necessary to establish the conditions for maintaining part of the conformational mobility and disorder of the melt or liquid crystal on cooling through the ordering transition that leads to positional and orientationally ordered molecules. Again, we expect the situation to be similar to the low molecular mass liquid crystals described in Sect. 5.2. In the main-chain liquid crystals discussed in Sect. 5.3.2 ordering of the mesogen enhanced the order in the flexible spacer considerably. Less effect is expected (and found) for the side-chain liquid crystal since tying of the spacers that experience on one end the ordering tendency of the mesogen to a continuous chain that tries to achieve a random coil seems more difficult. A statistical average of the entropy of transition to the isotropic state of over 100 liquid crystals [1] showed values of ΔS_i of 2 ± 1, 7 ± 5 and 15 ± 7 J/(K mol) for low molecular mass molecules, polymers with side-chain mesogen repeating units, and polymers with main-chain mesogen repeating units, respectively [1,11]. These values are to be compared with the entropy of transition for full orientational order of the mesogen alone of 20–50 J/(K mol) and the conformational entropy for gaining single bond mobility of 9.5 J/(K mol) (see Sect. 1).

The disordering transition at T_d was rarely analyzed quantitatively. The reported data are always too small for disordering from a well ordered crystal, but are usually not extrapolated to complete crystallization [1]. The existence of intermediate condis states is thus purely speculative. On cooling, it is frequent to find a direct transition of the liquid crystalline state to the LC glass (see Fig. 1.1). Placing the mesogen close to the backbone, as for example in poly(acryloyloxybenzoic acid),

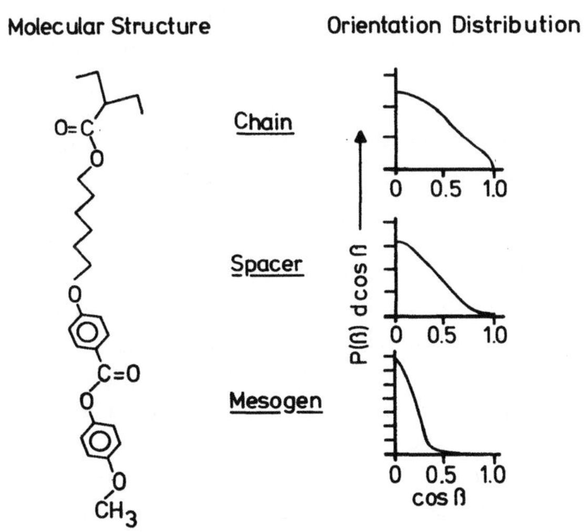

Fig. 5.11. Order distribution-functions of various parts of a polyacrylate with side-chain mesogen (phenylene-4-hydroxybenzoate) separated by a flexible spacer (hexamethylene). Determined from the angular dependence of the deuterium NMR line-shape. Presented by H. W. Spiess, Dechema Tagung 1987, see also Refs. [354,382]. The angle β is defined between the magnetic field and the corresponding molecular direction

seems to improve the crsystallization somewhat ($\Delta H_d = 22$ kJ/mol recalculated for 100% crystallinity) [379].

In the glassy state the liquid crystalline order of systems oriented in a magnetic field could be determined by deuterium NMR of specifically labelled compounds [354,380]. Figure 5.11 illustrates the decreasing order from mesogen to spacer to backbone chain. A local 180° flip-motion of the end phenylene group could be followed far below the glass transition temperature. All elements of condis crystallinity, conformational mobility and disorder are thus likely on cooling, but detailed studies of condis crystals of macromolecules with side-chain mesogens seem not to have been made as yet.

The special restriction caused by tying low molecular mass liquid crystalline substances to a polymer chain was also illustrated with amphiphilic liquid crystals [381]. A hexagonally close-packed structure of rod-like micelle cylinders of sodium 10-undecenoate with about 50% water lost during polymerization at 60 °C its structure and became isotropic. On cooling, a lamellar liquid crystalline structure, more suitable to accommodate the macromolecular backbone was found. Based on the discussions of Sect. 5.3.4 it is likely that with longer side-chain amphiphiles condis crystals could be grown in analogy to the soaps described in Sect. 5.2.3

5.3.4 Macroscopic Mesogens with Flexible Side-Chains

Three types of compounds with macroscopic mesogens have well-defined condis phases due to flexible side chains. Macroscopic mesogens of any shape have their glass transition temperature moved to so high temperatures that carbon-carbon bonds decompose thermally before a mobile mesophase is reached (see, for example, Fig. 5.1). A solvent is needed in such cases to stabilize a liquid crystalline phase (lyotropic liquid crystals). Instead of adding independent solvent molecules, it is, however, also possible to attach flexible side-chains to the mesogen and achieve thus an increase in entropy of fusion and, accordingly, a lowering of the melting temperature [382].

A large number of *inorganic layer crystals* such as micas, sodium silicates, niobate, uranate, vanadate, titanate, zirconium phosphate, graphitic acids, crystalline silicic acids, vanadium oxyhydrate, calcium phosphoric acid esters, and titanium disulfide develop alkyl crystals between their rigid crystal layers by ion exchange with, for example, alkyl ammonium salts and by intercalation [383,384]. The inorganic crystal layers provide a two-dimensional, macroscopic, rigid mesogen and more or less fix the position of the alkane chain. Structures of this type are known for some time [385,386]. A typical example is the clay beidellite, $Ca_{0.22}(H_2O)_x[Mg_{0.28}Al_{1.81}$-$(OH)_2Si_{3.56}Al_{0.44}O_{10}]$. The Ca^{++}-ions in the layer crystals were exchanged with n-tetradecylammonium chloride, and the remaining basal spaces were filled with more loosely held tetradecanol (or other alkanols) after removal of the water in high vacuum by intercalation. The organic structure was then $(C_{14}H_{29}NH_3^+)_{0.44}$-$(C_{14}H_{29}OH)_{1.56}[\,]$ [387]. X-ray diffraction revealed that the alkyl chains have a 0.23 nm^2 area available per molecule, that is somewhat more than in a typical paraffin lattice (0.19–0.20 nm^2); still, the alkyl chains are at low temperature in an all-trans conformation, standing perpendicular to the silicate sheets in bilayers [388]. On heating, sharp reductions in the layer separations by 0.12 ± 0.01 nm per step occur at 322, 333 and 345 K. These decreases in layer spacing have been explained by the successive introduction of 2g[1] kinks, i.e. gtg-conformations. Adiabatic calorimetry shows corresponding endotherms with a transition entropy of 4.8 J/(K mol) per alkyl chain for each of the transitions, values that are close to estimates for such kinks dynamically assembled in blocks [387] (see also the similarity in ΔS to the condis crystals of Fig. 5.7). At 367 K a more drastic reduction in layer-spacing occurs, coupled with a larger endotherm [383]. This was explained as a discontinuous increase in the kink concentration leading to gauche-block structures. Additional, smaller, stepwise decreases in layer spacing occur at higher temperatures (373, 383 K), indicating a remaining alkane crystal structure. Quantitative calorimetry is difficult in this region because of a slow loss of intercalated alkanol to compensate the larger, lateral space requirement of the conformationally disordered chains. While X-ray diffraction proves conformational disorder, calorimetry suggests dynamic disorder. Proof of high mobility, typical of condis crystallinity, was brought by proton NMR line-width and relaxation time measurement. The line-width, as measured by the second moment of the spectrum, decreases gradually from 20 G^2 at low temperature and shows a sharp jump to 1.0 G^2 at the 367 K transition [389]. The high G^2 values agree with $-CH_3$-mobility followed by introduction of mobile kinks, while the low values at higher temperature account for considerable mobility, but have by far not reached values typical for melts. Overall, these interesting compounds show much similarity to the amphiphiles described in Sect. 5.2.3.

Paraffin crystals held by a one-dimensional macroscopic rigid mesogen were observed for *poly(L-glutamate)s with long n-alkyl side chains* [390]

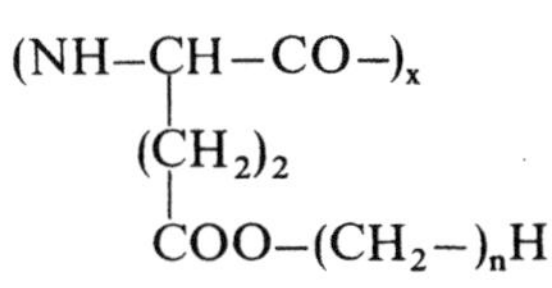

Above n = 8 increasingly sharp endotherms develop on heating the low temperature crystals. The short-chain substituted poly(L-glutamate)s ($n < 10$) show at low temperatures the common, hexagonal packing of rigid α-helices with one chain per unit cell. When precipitated from the typically cholesteric liquid-crystalline solutions, the solid films retain the liquid crystalline structure (LC glass, see Fig. 1.1). For poly(L-methyl-glutamate) ($n = 1$) the mobility of the side chains is only above 470 K sufficient to slowly lose the cholesteric twist of the α-helices with respect to each other [391]. Being tied to the rigid α-helix restricts the side-chain to local motion and the glass transition in the region of 470 K concerns the long-range motion of the molecule as a whole. With longer side-chains CH_2-sequences further removed from the rigid backbone can approach and crystallize. At that moment the α-helices lose their hexagonal packing and are placed instead at the surface of the interdigitated side-chain paraffin lamellae. It takes about 8 non-crystalline CH_2-groups to bridge from the α-helix to the paraffin crystal. At the endotherms that occur for $n = 10$, 12, 14, 16 and 18 at a disordering temperature of 248, 288, 314, 327, and 335 K, respectively, the paraffin diffraction pattern is lost and the α-helices are packed in a 2-dimensional hexagonal array again, but now with two chains per unit cell. Although the heat of transition per mole of crystallized CH_2 (about n-10) is close to 4 kJ, the typical paraffin heat of fusion, there is still another transition of almost negligible ΔH needed to reach the liquid crystalline (cholesteric) phase (303 to 335 K for n = 10 to 18, ΔH independent of n) [390]. It was suggested that the two chains per unit cell may form a so-called coil-coil structure. It remains a topic of debate if this high temperature state has enough translational and orientational restraint for the paraffin chain to be called a condis crystal. As indicated by the large entropy of transition, the molten side-chain segments are close to disorder in the melt. Typical condis crystals, as observed in paraffins (see Sect. 4.1) are perhaps expected with even larger side-chains so that the crystallized paraffin portions show conformational disorder within the temperature region of the existence of the lamellar crystals.

Similar *rigid chain macromolecules with long, flexible side-chains* have begun to be studied recently to achieve tractable liquid crystalline polymers with rigid backbone molecules. Depending on the restriction of the side-chain due to attachment to the rigid chain and on sufficient length to cause long-chain paraffin crystallization, condis crystallinity is expected. Polyimino[2,3,5,6-tetrakis(dodecyloxycarbonyl)-1,4-phenylene]-imino[2,5-bis(dodecyloxyterephthaloyl)] with 6 long side-chains per two-phenylene repeat was shown to have between 610 to 670 K a board-like liquid crystalline phase region and to be isotropic above 670 K. Below 610 K a large amount of dynamic disorder remains, but much further work is needed to elucidate the detailed condis structure [392].

A large series of poly(1,4-phenyleneterephthalate)s and poly(1,4-phenyleneterephthalamide)s with two alkoxides of up to 18 CH_2-groups each per repeating unit have been made [393–395]. For the terephthalates with more than 7 CH_2-groups per side-chain a new layered mesophase was observed that increased by 0.125 nm per additional CH_2—, indicative of an interdigitating arrangement of the side-chains normal to the rigid main-chains. Thermal analysis shows as many as three additional endotherms of various sizes below the transition temperature to the isotropic state that occurs from above 600 K to 465 K on increasing the number of side chain carbon atoms from 2 to 18. Parallel, the X-ray diffraction shows various degress of order and changes in the angle of side-chain to main-chain. As in the other examples of this section, the paraffin side-chains seem to assume changing degrees of dynamic disorder in their conformation, i.e. display different degrees of condis crystallinity at temperatures below the stability of the liquid crystalline phase.

5.4 Summary and Conclusions

Combinations of rigid and flexible segments complicate crystal structures considerably. In particular is was shown in Sect. 5.1 that seemingly rigid p-phenyl sequences may still show partial and complete ring-flips about the ring axes. The motion and introduction of defects based on ring-flips is much less cooperative than found for many flexible molecules.

The smallest members of the series of stiff rings (benzene, naphthalene, furan and substituted benzenes) are similar to the small cycloalkanes (Sect. 3.1) and are capable of beginning jump-motion between symmetry related positions without thermal transition. Destroying the symmetry causes similar motion to be connected with entropy-gain in a first-order transition (furan compared to benzene). With too little difference in interaction-energy, ordering of disordered molecules may become impossible, leading on cooling to glasses that have a non-zero entropy at zero kelvin.

Low molecular mass liquid crystals with a rigid mesogen could be shown to display frequently additional condis phases, particularly if long, flexible paraffin-chains are part of the molecule. Three types of conformational disorder are frequent. Geometrical disorder due to statistical placement of molecules that have only slightly asymmetric shapes, partial or full flips of p-phenylene groups, and the flexible chain disorder described in Sect. 4. Often the gain (or loss) of conformational disorder occurs in several steps. Quite often the sum of disorder entropy over all transition is less than expected. This may be explained either through the existence of additional transitions at lower temperature or some continuous, non-cooperative ordering (or disordering) as suggested in Sect. 2.

The problem of identifying phases in the area between highly-ordered, smetic liquid crystals and condis crystals is well recognized and often only solvable when a complete motional, structural, and entropic analysis is available.

Amphiphilic molecules represent a vast resource for condis crystals. It is easy to establish long-range order in one of the two components of the molecule that can tolerate higher temperatures than the crystal of the other component. At higher temperature one component can then become mobile and conformationally disordered before the other, but is hindered to take on orientational and positional disorder due to attachment to the thermally more stable crystal. The waxy phases of soaps, many high temperature phases of long chain carboxylates and substituted ammonium salts as well as lipids are representative of such condis crystals.

Semiperfluoroated paraffins have special condis phases in which both halves of the molecule have their own dynamic conformational disorder.

Connecting rigid structures to flexible chain segments produces very special macromolecules. Their properties can be varied over a wide spectrum to yield from rubbery (segmented polyurethanes) to high modulus (main chain mesogen) fibers. For many of these polymers that have seen attention only recently, the analysis of conformational disorder is not yet complete. Dynamic conformational disorder is of importance in flexible chain segments as well as in mesogens with ring mobility (p-phenylene) or flexibility (cyclosiloxanes).

Mesogens in the side-chain have not yet lead to clearly identified condis states of the flexible spacers, probably because of the need of longer spacers. Usually the liquid crystalline state changes on cooling directly to the LC glass.

Macroscopic mesogens coupled to flexible side-chains, finally, represent a group of interesting compounds that can be used as model systems for membranes for the illustration of self-solubilization of rigid chain molecules, and also as models for the study of condis crystals in general. Questions of the length of flexible spacers needed for crystallization, the influence of changed lattice spacing on defect mobility and production, and solubilization as a function of side-chain length could be approached.

Condis crystallinity could thus be shown to bring new insight to a large variety of small and large molecular mass compounds.

6 Condis Crystals of Stiff Macromolecules

The main industrial interest in linear macromolecules centered for many years around their "plastic" behavior, i.e. the ease of heating the material above its glass or melting transition and shaping it in the liquid state by extrusion, injection, drawing, molding or blowing, to be followed by resolidifying on cooling. The reason for this convenient behavior is the flexibility of the molecular chain that results from rotation about the backbone bonds. The less flexible the molecule, the higher the glass transition or melting temperature, and the harder it is to use the material as a "plastic". Really stiff molecules remained for many years laboratory curiosities as so-called "brick-dust", powdery crystalline or amorphous substances that could not be brought to the high temperatures needed for the liquid state, often because of intervening decomposition.

6.1 Chain Extension and Flexibility

The high flexibility of the molecules of "plastics" has for certain applications the disadvantage that one cannot make use of the potentially high modulus and strength of polymeric materials because of difficulties in orienting and extending flexible chains. Melts of flexible, linear macromolecules consist of interpenetrating random coils with end-to-end distances proportional to the square root of the chain length. A typical polyethylene of 280,000 molecular mass, has, for example, in the melt an end-to-end distance of about 53 nm which compares to the all-trans, extended zig-zag chain-length of 2,530 nm. On crystallization such molecules grow lamellae with a fold-length of also typically 20 to 50 nm [47,155]. To go to full extension from such melt or crystal one needs thus mechanical drawing without molecular slip by a factor 50 to 100. For many years this was commercially impossible. Stiff chain molecules, as, for example, polyparaphenylene, discussed in Sect. 5.1.4 as the limit of a macroscopic liquid-crystal mesogen, have, in contrast, a natural tendency to arrange themselves parallel and should thus be ideal to produce high modulus fibers. Their high glass or melting transitions must, however, be overcome to achieve such parallel packing.

Several very special methods have been developed to overcome the problems of sufficient extension of flexible linear macromolecules. One method is the crystallization during polymerization where the random, polymerized phase is avoided [155,396]. Another, the crystallization into a suitable condis state as discussed, for example, for polyethylene and polytetrafluoroethylene in Sect. 4.1 and 2, where within the condis phase the molecule has enough mobility to permit extension. Several other techniques accomplishing the substantial extensions needed to approach the theoretical modulus involve controlled reduction of entanglements of molecules of properly chosen molecular mass by superdrawing, growth from flowing solutions, pressure extrusion and gel spinning [397,398]. Moduli of 100 GPa and tensile strength of 2–3 GPa have been achieved for such flexible molecules as polyethylene. Larger scale commercial application of these techniques seems to be in the near future.

The problem of high transition temperatures in rigid, strictly linear macromolecules is also difficult to overcome. The obvious solution to find a solvent

Table 6.1. Rigid rod segments with parallel rotation axes

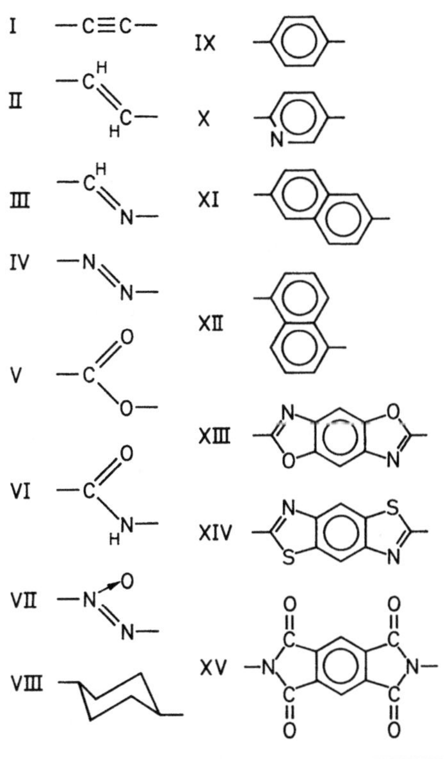

is often difficult for macromolecules such as, for example, represented by the repeating units I, IX, X, XIII, XIV and XV of Table 6.1 or their combinations. A second approach is the production of the stiff chains by chemical reaction of the monomer or a flexible precursor after or during production of the required shape (fiber, film, surface layer or mold). Conversion may in some cases, however, be limited. Two examples are polyimides, such as IX + XV from the corresponding polyamic acids and polybenzobisthiazoles IX+XIV from 2,5-diamino-1,4-benzenedithiol and terephthalic acid [399,400].

A next step to handle such molecules is to relax the requirement of strictly linear molecules to macromolecules made of rigid-rod segments with parallel, but off-set rotation axes. Such segments may introduce a limited degree of disorder and mobility, the so-called crank-shaft structure and motion. Some examples are listed in Table 6.1 (II–VIII and XI–XII). The commercial development of high modulus fibers of poly(p-phenylene terephthalamide) in 1965 ignited much of the present-day interest in this type of stiff molecules. References 399 and 400 give a general review of the topic. In a suitable solvent such molecules of limited mobility may form a lyotropic liquid crystalline state that orders easily on flow through a spinneret, for example. On precipitation the polymer becomes, depending on conditions, the corresponding fully or partially glassy mesophase [trade name of the du Pont fiber: Kevlar; structure IX+VI(reverse)+IX+VI, see Table 6.1; solvent, conc. sulfuric acid]. Later or simultaneous annealing can introduce high degrees of crystallinity. Similarly constructed polyesters (V), stilbens (II), azoxy (VII) or azo (IV) compounds with properly chosen ring compounds (VIII to XII) may even have sufficiently reduced transition temperatures to be processible directly from an anisotropic melt. Estimates of the size of the statistical segment (Kuhn segment) for a random structure of such crank-shaft polymers are rather large, about 50 to 200 nm [401]. This compares to less than one nanometer for the most flexible macromolecules such as polyethylenes or 2–5 nm for aromatic polyamides where the segmental rotation axes are not parallel [i.e. poly(m-phenylene isophthalamide)]. Introduction of non-parallel rotation axes cause thus considerably more disorder into the polymer structure than the crank-shaft structure. A final step of reduction of the solid-state transition temperatures of polymer chains is the introduction of fully flexible spacers into the main-chain, as discussed in Sect. 5.3.2, or the attachment of side-chains. The latter maintain the main-chain stiffness, as discussed already in Sect. 5.3.4. Both methods can lead to a sufficient reduction in glass or melting transition temperature to produce suitable "plastic", and by judicious choice of type, size and number of the flexible main-chain spacers or side chains, it is possible to maintain the liquid crystalline tendency of the molecule. It is thus possible to produce high performance fibers tailored to specific needs.

The question of condis crystallinity has not been analyzed to any detail for rigid backbone molecules. As outlined in Sects. 4 and 5 there is, however, a great potential for such mesophases. The flexible spacers or side-chains are likely, under proper conditions, to form the paraffin or polyethylene-like condis phases. Their existence-range is being broadened due to the mismatch of crystal structure of flexible and rigid molecule segments, and the increase in melting temperatures is caused by the stiff segments. The rigid molecule portions, on the other hand, may show the entropically less effective small-angle and 180° flips of phenylene groups in a condis state (see Fig. 5.1). In addition, one may speculate about dynamic disorder of the crank-shaft structures.

In this next sections an attempt will be made to collect on the examples of poly(p-phenylenebenzobisthiazole), poly-p-oxybenzoate and poly(p-phenylene terephthalamide) the sparse evidence suggesting that dynamic disorder may sometimes also be of importance for the understanding of stiff molecules.

6.2 Strictly Linear Macromolecules

Strictly linear macromolecules would be, for example I, IX, X, and XIII to XV of Table 6.1. Of these not all have been made or are chemically stable and little is known about their structure. *Polyparaphenylene*, as one example, was described in Sect. 5.1.4 and linked to the better known oligomers (Sect. 5.1.1–5.1.3). The LC glass state proposed for the polymer would be difficult to dissolve into a lyotropic liquid crystal because of the seeming lack of high temperature solvents that could reduce the glass transition temperature sufficiently. By itself decomposition occurs before the glass transition temperature is reached.

More successful was the handling of *poly(p-phenylene benzobisthiazole)* (PPB)

(IX+XIV) as an example of a combination of paris of suitable straight-chain segments from the above list. The knowledge about this system has been summarized in Ref. [399]. Fibers have been produced from a liquid crystalline spinning dope of 5–6% polymer in poly(phosphoric acid) $(HPO_3)_x$. Heat treated fibers reached moduli of up to 250 GPa with tensile strength of about 2.3 GPa and 1.5% extension at break. The structure in the solid state is that of a nematic arrangement of the stiff chains in a hexagonal array with varying lateral and longitudinal alignment between the molecules. A common problem in all such solution-precipitated, stiff-chain molecules seems the inclusion of voids in the structure. Again, the structure may be describable as an LC glass, but probably with varying amounts of crystallinity. The crystallinity of only slightly more flexible molecules has been better analyzed for poly(p-phenylene terephthalamide) (see Sect. 6.4).

6.3 Polyoxybenzoate and its Copolymers

The homopolymer *poly(4-oxybenzoate)* (POB, structure IX+V of Table 6.1) is an intractable polymer. But by crystallization during polymerization [396] POB can be grown into extended chain crystals [402]. The chain conformation is shown in Fig. 6.1. Lamellar oligomer crystals grow from several solvents in stacks and can, above 700 K, be brought to interconnect to typical extended chain lamellae [47] of degrees of polymerization of 200 and more [402]. Similar crystal extension mechanisms have been observed for poly(ethylene terephthalate) and selenium, and to some measure also for poly(oxymethylene) [155]. The crystallinity of such samples is high. An orthorhombic crystal structure $(P2_12_12_1)$ is likely, with the molecular chains following a 2_1 screw axis, i.e. a regular crank-shaft structure with antiparallel chains [403], similar to the poly(4-benzamide) structure [245]. Two different packings with identical chain-repeat (1.25 nm) exist at room temperature (polymorphs I and II), but both transform at 560–620 K partially reversibly into a polymorph III that has been called either a liquid crystal [405] or a condis crystal [1].

The polymorph III of p-POB has been further analyzed on its 95/6 p/m copolymer

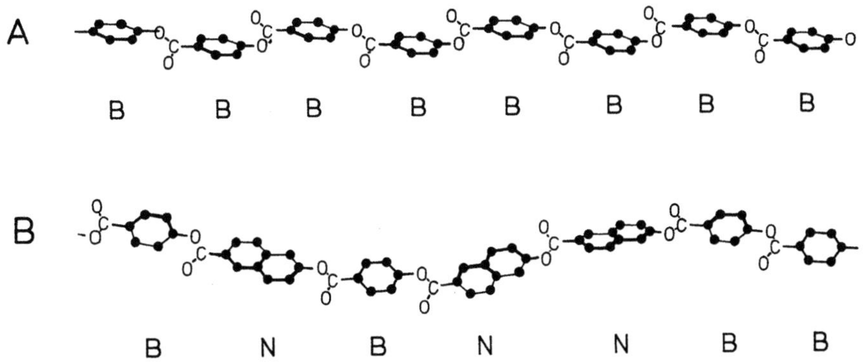

Fig. 6.1. Chain conformations in crank-shaft like polymers. A: The regular structure of poly(p-oxybenzoate); B: random sequence of POB/ON

that shows sufficient mobility and thermal stability to be oriented to fibers [404]. Little difference exists for the mesophases of homo- and copolymer. They have an approximately 11% lower density and a hexagonal lateral packing with an only slighly shortened chain-axis repeat (1.23 nm) [403]. Heating the (copolymer) sample to 750 K, the equatorial X-ray reflections become increasingly broad and diffuse, approaching liquid-like packing between the chains, but with remaining sharp meridional X-ray reflections (chain-repeat at 743 K, 1.22 nm) [404]. At 753 K a disorientation of the copolymer chains to the isotropic state takes place, a transition that could not be observed for the homopolymer because of decomposition. The sharp 00l-reflections of polymorph III make it unlikely that it possesses much translational motion and positional disorder. A condis crystal with increasing conformational disorder and mobility is thus a better description of the mesophase of p-POB in the 600–700 K temperature range. An increase in the m-component in the copolymer to 50/50 yields crystal structures that are still similar to the homopolymer low-temperature phase with laterally disturbed packing in the crystallographic b-direction and, again, a transition into the condis phase at 553 K. At 673 K a true liquid crystalline structure has been reported that becomes isotropic at 713 K [404].

Even more mobility was introduced into POB *by copolymerization with ethylene terephthalate* (ET). The ET disturbs the stiff chain considerably more than an occasional off-axis meta-bond discussed above. A typical copolymer system with repeating-unit isodimorphism [47] has been reported [405] although the early work [406] seemed to indicate little crystallization in the middle concentration range. Poly(ethylene terephthalate) (PET) is a typical flexible macromolecule without mesophase transitions (but see also Sect. 5.3). Beginning with 25–30% OB the copolymers form anisotropic melts. The glass transition is surprisingly insensitive to oxybenzoate content and jumps ultimately without intermediate states from about 350 K, characteristic of PET, to 440 K, characteristic of POB [405,407]. The intermediate range shows often two-phase samples with both glass transitions. The ET-rich copolymers show crystallinity related to the PET crystal structure [406], while the OB-rich copolymers are similar to the POB crystal structure [408]. The crystal form of the high oxybenzoate samples at room temperature is, however, that of the condis phase [408]. The high temperature POB crystal may thus tolerate the ET defects to some degree, but the defects prevent the large-scale formation of the more compact, low temperature crystal forms. The continuous change in peak temperature of the melting endotherms with OB concentration from 500 to 600 K seems to indicate, however, a certain amount of crystallinity of small extension may still be present at low temperature [405]. The situation is thus not fully clear at present. It may be suggested that the low glass transition temperature results from the amorphous phase at low OB content that changes to a liquid-crystalline-like phase without major change in glass transition. The high glass transition temperature results then from a condis-crystalline POB form III phase. The observed endotherms, finally, may be small crystallites, the result of isodimorphism of ET and OB repeating units.

Even more interest has been shown for the *copolymers of p-oxybenzoate with 2-oxy-6-naphthoate* (structure: XI + V, Table 6.1 POB/ON) because of their potential commercial uses [409]. The glass transition changes over the whole composition range from 434 K for POB to 377 K for 58/42 POB/ON to 399 K for PON with a normal jump in heat capacity throughout the concentration range [31–34 J/(K mol)], but with an enormous transition breadth in the ON-rich copolymers (about 200 K)[412]. Without special heat treatment, the observed endotherms are rather small (0.4 to 4 kJ/mol) and follow, as POB/ET, a melting peak-temperature diagram with a minimum (isodimorphism, minimum endotherm temperature at about 60 mol-% OB at 530 K)[410]. A detailed crystallization and melting study revealed that even close to the minimum melting temperature high crystallinity (about 60%) can be achieved on long-term annealing[411]. On cooling, fast ordering to a low degree was observed, followed by an independent, logarithmically time-dependent, slow crystallization (second melting peak) which increases in perfection with time and temperature. At high temperature (above the endotherm) X-ray diffraction shows only broad, melt-like diffraction on the equator, while on the meridian several sharp arcs remain. The latter were identified on all related copolymers as aperiodic, parallel sequences of the different repeating units[412,413]. In fact, the meridional reflections could be used to prove the absence of blockiness of the copolymer. Figure 6.1 illustrates a typical random sequence of POB/ON of almost linear extension.

Special models have been developed for the crystallization of these and related stiff-chain aromatic copolymers[414,415]. The basic crystal was suggested to be of the condis crystal type with rotational disorder based on conformational mobility, proven by mechanical and dielectric relaxation[414]. The morphology of the superimposed crystals was proposed to be based on a matching of order between adjacent molecules[415] similar to "cold crystallization" originally proposed for the crystallization of PET-copolymers from the glassy state[416].

6.4 Aromatic Polyamides

The two early developed crank-shaft-type, stiff macromolecules are *poly-p-benzamide*[417] ("Fiber B" structure IX+VI) and *poly(p-phenylene terephthalamide)*[418] [Trade names Kevlar, Arenka, PPT, structure IX+V (reverse)+IX+VI, see Table 6.1]. The latter has seen major industrial production as high modulus fiber. The fibers are spun at 70 to 90 °C from a liquid-crystalline (nematic) spinning dope of concentrated sulfuric acid (99.8%) with about 20% PPT into water below 280 K. The as-spun fibers are already highly oriented, but are usually heat treated briefly at 750 to 800 K in an inert atmosphere. Typical moduli of the commercial products are 50 to 150 GPa with tensile strengths of 2–3 GPa with 2–4% elongation at break. The thus produced fibers begin to decompose in air in excess of about 700 K without prior fusion[399]. Scanning calorimetry seems to show a first major endotherm starting at about 825 K[419], but the transition type has not been identified.

The precipitation process is assumed to lead either from the lyotropic, nematic liquid crystalline state via precipitation with water under maintenance of ordered water-polymer layers to the crystal form II, or with other solvents (and also from low polymer concentrations with water) through disordered solvent-polymer structures to crystal form I[420]. In both crystal structures the ultimate polymer crystal layers are H-bonded in the crystallographic bc-plane (100), as shown in Fig. 6.2. In crystal form I the second molecular chain goes through the center of the unit cell (Pn or $P2_1/n$ space group, 2 chains per unit cell, monoclinic, pseudo-orthorhombic)

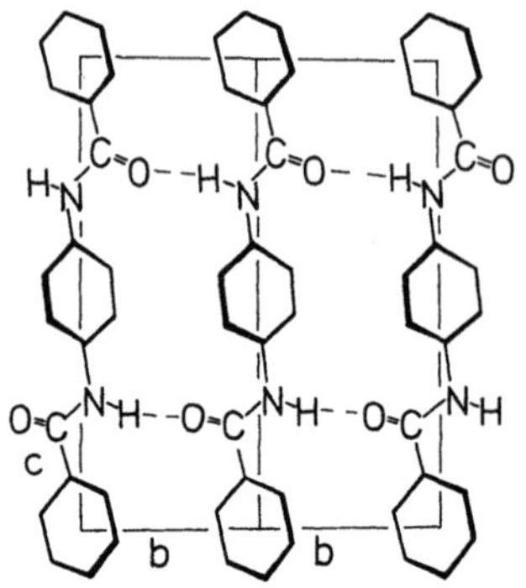

Fig. 6.2. Hydrogen bonded sheet of poly(p-phenylene terephthalamide). Pseudoorthorhombic crystal structure. Two chains per unit cell. For crystal form I the second chain centers the ab-plane. [a = 0.787 nm, b = 0.518 nm, c = 1.29 nm, β = 90°]. Rotation angles of the phenylene/amide planes 38° and of the terephthalate/amide planes −30°

[421]. In crystal form II the (100) layers are stacked on top of each other [420]. The overall morphology of PPT fibers has been described as practically fully crystalline, but with periodic defect layers and ±5° pleating of the fibrils making up the macrofiber. Etching experiments yield 35 nm long single crystal grains, but with continued correlation of the stiff chains across defect layers to about 80 nm [422].

The morphology and structure of PPT crystals grown directly from H_2SO_4 solution by solvent evaporation was somewhat different [423]. Pyramidal single crystals of 5 to 10,000 nm were observed, made up of rectangular platelets of molecular-mass segregated single crystals [50–200 nm (chain direction) by up to 1000 nm]. The lattice parameters in these single crystals are different from both, forms I and II. Spherulites (negative) grown in the same way consist of radiating lamellae composed of crystallites interconnected in the radial direction, with the extended chain direction tangential to the spherulite.

All these morphological and structural studies seem to indicate that from the nematic state in solution, with dynamic conformational and positional disorder, rather high three-dimensional order is produced with little change of remaining conformational mobility in the crystalline regions. Solid state nuclear magnetic resonance studies (^1H and ^{13}C) [424] indicate no large scale conformational motion excluding even phenylene rotations of more than 20° up to about 500 K.

The PPT structures, in contrast to the polyesters, seem thus to be best described as crystals with (minor) defect regions. Of interest would be the introduction of structural disorder that may permit more large-scale motion. It must be considered, however, that even in dilute solution, PPT shows no evidence of cis-isomeric amide groups [424], i.e. the only types of motion seem, indeed, crank-shaft motion and local phenylene rotations.

6.5 Summary and Conclusions

Strictly linear macromolecules such as poly(p-phenylene) cannot be brought into the molten state, and thus have solid state structure set by the polymerization condition. Conformational mobility is usually limited to ring rotation. Little details are known about these polymers. For dissolution, solvents with strong interactions and of high temperature stability are needed, as for example found for poly(p-phenylenebenzobisthiazole). The precipitated structures have of necessity extended polymer molecules, but little seems to be known about disorder and mobility. In both examples treated in Sect. 6.2 an L.C. glass structure with unknown crystallinity by 3-dimensional registry over limited volumes seems to be the present picture.

Limited reduction of strict linearity by introduction of crank-shaft disorder and motion allows a broader variety of behavior as is discussed in Sects 6.3 and 6.4. The crystallization and solid state behavior of these still rather stiff molecules is basically different from flexible molecules. Crystallization of flexible macromolecules is governed by their random coil macroconformations, disentanglement between molecules, sequential repeating unit addition to a crystal surface, and chain folding. Crystallization of stiff macromolecules occurs, in contrast, via aggregation from the lyotropic liquid-crystalline or anisotropic molten state where chains are already extended, parallel, free of entanglements and in proper sequence for crystallization. The remaining problems are in this case rotational (conformational) adjustment and translational motion into register on a repeating-unit scale and, for ultimate perfection, into register on a molecular scale (matching of molecular and crystal dimension). Depending on the degree of perfection reached, the resulting solid is an L.C. glass (no perfection beyond the liquid crystalline structure), or has partial or complete crystallinity. Once crystallized, an equilibrium, dynamic conformational disorder may exist at sufficiently high temperature as proven for the poly(p-oxybenzoates).

Even more reduction of linearity and introduction of flexibility leads to the molecules described in Sect. 5 and ultimately to full flexibility as described in Sect. 4. Condis states can be found in all these types of molcules.

7 Conclusions

Over 100 substances with condis phases have been identified. About 2/3 of these were small molecules, the rest macromolecules. These substances come from the fields of organic, inorganic and biological chemistry documenting the universal need to understand dynamic conformational disorder.

In the process of identification of condis crystals it was observed that conformational mobility alone is not sufficient to prove the presence of a condis phase. Large amplitude molecular jump motion may be possible already in crystals without disorder if the symmetry is identical before and after the jump. The frequencies of these jumps can be surprisingly large and the moving parts of the molecules substantial. In the condis phase quick reptation can lead to extension of folded chain crystals, and is possibly also involved in rearrangements on mechanical deformation and membrane functions.

Directive driving forces such as reduction in surface free energy of folded chain lamella or mechanical energy exerted on drawing may lead in the condis state to substantial structure changes. The formation of extended chain crystals is most interesting since it can lead to high modulus and strength macroconformations. The special rotoreptation motion in cycloalkanes opens a chance at quantitative determination of the appropriate diffusion kinetics.

Liquid crystals can be made from rigid mesogens of various sizes and shapes by attachment of conformationally disordered, flexible chains. Much of the lower order achieved on cooling liquid crystals to the solid state, in contrast to materials not having a liquid crystalline phase, finds its explanation in the existence of condis crystals or CD glasses.

The large variety of conformational motion is reflected in the different degrees of cooperativity in the introduction of disorder that may range from gradual to sharp first order transitions. The frequent freezing of motion in macromolecules to glassy phases instead of ordering is also the reason for the frequently observed non-equilibrium states, i.e. it is possible to have multiphase structures even for one-component systems. The two-phase model of semicrystalline polymers must be extended in the presence of condis phases to a three-phase model (crystal, condis crystal or CD glass, melt or glass).

The differences between the three mesophases, liquid crystal, plastic crystal and condis crystal could be clarified and, although difficult to classify, intermediate cases remain. If the molecule permits dynamic conformational disorder in the liquid phase, the same is usually also possible in a similar temperature range in the liquid-crystalline and plastic-crystalline phases.

Condis crystals are so broadly distributed that their further exploration is expected to lead to substantial progress in the understanding of many types of materials.

8 Acknowledgements

Much of this work was supported at RPI by the Polymers Program of the US National Science Foundation (Grant Number DMR 83-17097) and at the U. Freiburg by the Deutsche Forschungsgesellschaft within the SFB 60. The cooperation, discussion and ultimate writing of this paper was generously helped by the Alexander von Humboldt Stiftung through a Senior US Scientist Award (B. W., 1987). One of us (J. G.) was supported during part of his work on this review by the Max Planck Institut für Polymer Forschung in Mainz. Special thanks go to Prof. E. W. Fischer for support during this time.

9 References

1. Wunderlich B, Grebowicz J (1984) Thermotropic mesophases and mesophase transitions of linear, flexible macromolecules. Adv. Polymer Sci., *60/61*: 1
2. see for example: Sherwood N (ed) "The Plastically Crystalline State. (Orientationally-disordered crystals)." J. Wiley and Sons, Chichester 1979
3. see for example: Gray GW "Molecular Structure and the Properties of Liquid Crystals." Academic Press, New York 1962
4. Ehrenfest P (1933) Phase changes classified according to the singularities of the thermodynamic potential. Acad. Sci., Amsterdam *36*: 153 Suppl. 75b, Mitt. Kammerlingh Onnes Inst., Leiden
5. Richards JW (1897) Relations between the melting-points and the latent heats of fusion of the metals. Chem. News *75*: 278
6. Walden P (1908) Über die Schmelzwärme, spezifische Kohäsion und Molekulargröße bei der Schmelztemperatur. Z. Elektrochem. *14*: 713
7. Wunderlich B "Macromolecular Physics, Vol. 3, Crystal Melting." Academic Press, New York, 1980

8. Vorlaender D (1933) Remarks on Liquocrystalline resins and laquers. Trans. Farad Soc. *29*: 207
9. Adachi K, Suga H, Seki S (1970) The glassy crystalline state — a non-equilibrium state of plastic crystals. Bull. Chem. Soc. Japan *43*: 1916
10. Wunderlich B (1960) Study of the change in specific heat of monomeric and polymeric glasses during the transition. J. Phys. Chem. *64*: 1052 The upper limit in ΔC_p applies for larger mobile parts in molecules, see Wunderlich B, Gaur U Pure Appl. Chem. *52*: 445 (1980)
11. Wunderlich B, Möller M, Wiedemann HG (1986) Condis crystals of small molecules I. The concept and limitations. Mol. Cryst. Liq. Cryst., *140*: 211 Wiedemann HG, Grebowicz J, Wunderlich B II. The polymorphs of N,N′-bis(4-n-octyloxybenzal)-1,4-phenylenediamine (OOPBD), ibid. 219; Cao M, Wesson J, Loufakis K, Möller M, Wunderlich B III. Thermal analysis of plastic and condis crystals of some cyclosilanes, ibid. 231; Möller M, Wunderlich B IV solid state NMR of some cyclosilanes, ibid. to be published (1988/89)
12. see for example: Ubbelohde AR "Melting and Crystal Structure." Oxford Univ. Press (Clarendon) London 1965; and "The Molten State of Matter, Melting and Crystal Structure." Wiley, New York, 1978
13. Lennard-Jones JE, Devonshire AF (1939) Critical and cooperative phenomena. III. A theory of melting and the structure of liquids; IV. A theory of disorder in solids and liquids and the process of melting. Proc. Roy. Soc. London, Ser. A. *169*: 317; *170*: 464 See also deBoer J (1952) Theories of the Liquid State. Proc. Roy. Soc. London, Ser. A *215*: 1
14. Hirschfelder J, Stevenson D, Eyring H (1937) A theory of liquid structure. J. Chem. Phys. *5*: 896
15. Flory PJ "Structural Mechanics of Chain Molecules." Wiley (Interscience), New York, 1969
16. Birshtein TM, Ptitsyn OB "Conformations of Macromolecules" (Translated from the Russian by SN Timasheff and MJ Timasheff). Wiley (Interscience), New York, 1966
17. Volkenstein MV "Configurational Statistics of Polymeric Chains." (Translated from the Russian by MJ Timasheff and MJ Timasheff) Wiley (Interscience), New York, 1963
18. Grossmann HP, Pechhold WR (1986) The high temperature transition of polyethylene into a condis phase, described on the basis of the cluster-entropy hypothesis (CEH). Colloid and Polymer Sci., *264*: 415
19. Pechhold W, Liska E, Grossmann HP, Hägele PC (1976) On present theories of the condensed polymer state. Pure Appl. Chem. *46*: 127
20. Pechhold W, Grossmann HP, v Soden W (1982) Versetzungstheorie des Schmelzens auf der Grundlage der Cluster Entropie-Hypothese (CEH). Colloid and Polym. Sci. *260*: 248
21. Einstein A (1907) Die Plancksche Theorie der Strahlung und die Theorie der spezifischen Wärme. Ann. Physik, *22*: 180 for correction see Ann. Physik *22*: 800 (1907)
22. Debye P (1912) Zur Theorie der spezifischen Wärmen. Ann. Physik, *39*: 789
23. Tarasov VV (1945) The theory of the specific heat of high polymers. Compt. Rend. Acad. Sci. URSS *46*: 110 *54*: 795 (1946); see also Zh. Fiz. Khim. *24*: 111 (1950); *27*: 1430 (1953)
24. Wunderlich B (1962) Motion in polyethylene. II. Vibrations in crystalline polyethylene. J. Chem. Phys. *37*: 1207
25. Wunderlich B, Baur H (1970) Heat capacities of linear high polymers. Adv. Polymer Sci., *7*: 151
26. for a recent summary see: Wunderlich B, Cheng SZD (1986) Thermal Analysis of Linear Macromolecules. Gazz. Chim. Italiana, *116*: 345
27. for an overall assessment see: Bu HS, Cheng SZD, Wunderlich B (1987) Study of skeletal vibrations used to compute heat capacities of solid macromolecules. J. Phys. Chem., *91*: 4179
28. Judovits L, Bopp RC, Gaur U, Wunderlich B (1986) Heat capacities of polystyrene, substituted polystyrenes and crosslinked polystyrenes. J. Polymer Sci., Polymer Phys. Ed. *24*: 2725
29. Wunderlich B (1962) Motion in polyethylene. III. The amorphous polymer. J. Chem. Phys. *37*: 2429 see also J. Polymer Sci., Part C, *1*: 41 (1963)
30. Loufakis K, Wunderlich B The computation of heat capacities and transition parameters of liquid macromolecules based on a statistical mechanical approximation. J. Phys. Chem., to be published Vol. *92(3)*, 1988
31. Grebowicz J, Aycock W, Wunderlich B (1986) Heat capacities of 1,4-polybutadienes. Polymer, *27*: 575

32. see for example: Fyfe CA "Solid State NMR for Chemists." CFC Press, Guelph, 1983
33. Darmon I, Brot C (1967) The orientational freedom of molecules in crystals I. Joint interpretation of thermal and NMR data in an "all or nothing" correlation of orientation theory. Mol. Cryst. *2*: 301 see also Brot C, Lassier-Govers B Ber. Bunsen Ges. *80*: 31 (1976)
34. Kilpatrick JE, Pitzer KS, Spitzer R (1947) The thermodynamics and molecular structure of cyclopentane. J. Am. Chem. Soc. *69*: 2483 see also Pitzer KE, Donath WE J. Am. Chem. Soc. *81*: 3213 (1959)
35. see for example the reviews: Strauss HL Pseudorotation (1983) A large amplitude molecular motion. Ann. Rev. Phys. Chem., *34*: 301 Carriera LA, Lord RC, Mallory Jr TB (1979) Low frequency vibrations in small ring molecules. Top Curr. Chem., *82*: 1
Legon AC (1980) Equilibrium conformations of four and five-member cyclic molecules in the gas phase: Determination and classification. Chem. Rev. 231
36. DeGennes PG (1971) Reptation of a polymer chain in the presence of fixed obstacles. J. Chem. Phys. *55*: 572
37. Baur H "Einführung in die Thermodynamik der irreversiblen Prozesse", Wissenschaftliche Buchgesellschaft, Darmstadt 1984
38. Allen G (Ed) "Molecular Structure and Properties", Butterworths, London 1972
39. Hedvig P "Dielectric Spectroscopy of Polymers", Adam Hilger Ltd., Bristol 1977
40. Abragam A "The Principles of Nuclear Magnetisme", Oxford University Press, London 1961
41. Ward IM "Mechanical Properties of Solid Polymers", Wiley-Interscience Inc., New York 1971
42. Davies RO "Sur les soit-disant transitions de second ordre dans les milieux a relaxation", Comptes Rendus IUP, Société de Chimie Physique, Paris 1952
43. Davies RO, Jones GO (1953) Thermodynamic and Kinetic Properties of Glasses, Adv. Phys. *2*: 370
44. Münster A "Principles of Statistical Mechanics" in Encyclopedia of Physics (Ed. S. Flügge) Vol. III/2, Springer-Verlag, Berlin and New York 1959
45. Frenkel JJ "Kinetic Theory of Liquids" Oxford University Press, London 1946; reprinted by Dover, New York 1955
46. Parsonage NG, Stavely LAK "Disorder in Crystals." Clarendon Press, Oxford, 1978
47. Wunderlich B "Macromolecular Physics, Crystal Structure. Morphology, Defects", Academic Press, New York, 1973
48. Blasenbrey S. Pechhold W (1967) Eine neue Theorie zur Rotationsumwandlung der n-Paraffine (Kinkentheorie), Rheol. Acta *6*: 174
49. McMahon PE, McCullough RL, Schlegel AA (1967) Molecular Mechanics of Point Defects in Polyethylene, J. Appl. Phys. *38*: 4123
50. Baur H (1974) Bemerkungen zur Defekttheorie von n-Paraffinen und Polymeren, Colloid & Polymer Sci. *252*: 641
51. Baur H "Zur Theorie der Umwandlungserscheinungen in n-Alkyl-Lamellen", Habilitationsschrift DB 1992 + a, Universität Hannover 1977
52. Baur H (1974) Zur Dfekttheorie des einphasigen Vorschmelzens von n-Paraffinen; Teil I: Bragg-Williams-Näherung", Colloid & Polymer Sci. *252*: 899
53. Baur H (1976) Wärmekapazität einer Mischung von rotationsisomeren Ketenmolekülen im Gleichgewicht, Colloid & Polymer Sci. *254*: 217
54. Baur H (1975) Zur Defekttheorie des einphasigen Vorschmelzens von n-Paraffinen; Teil II: Quasi-chemische Näherung, Lagaly-Fitz-Weiss-Effekt, Progr. Colloid & Polymer Sci. *58*: 1
55. Egan CJ, Kemp JD (1937) Ethylene, the heat capacity from 15 K to the boiling point. The heats of fusion and vaporization. Vapor pressure of the liquid. The entropy from thermal measurements compared with the entropy from spectroscopic data. J. Am. Chem. Soc. 59: 1264
56. Brecher C, Halford RS (1961) Motion in molecules in condensed systems. XI. Infrared spectrum and structure of a single crystal of ethylene. J. Chem. Phys. *35*: 1109
57. van Ness GJH, Vos A (1979) Single-crystal structures and electron density distributions of ethane, ethylene and acetylene. III. Single crystal X-ray structure determination of ethylene at 85 K. Acta Cryst. *B35*: 2593
58. Trappeniers NJ, Ligthart FAS (1973) Proton spin relaxation in ethylene. A new high pressure phase. Chem. Phys. Lett. *19*: 465
59. Page JL, Rushworth FA (1969) Nuclear spin-lattice relaxation in ethylene and trans-dichloro-

ethylene, J. Phys. C. *2*: 415 see also Van Putte K, Van den Enden JJ Phys. C. *4*: L163 (1971) and Hoch MJR, Rushworth FA Phys. Lett. *12*: 292 (1964)
60. Bates JB (1973) Raman spectra of crystalline cyclopropane, cyclopropane-d_6 and C_3H_6—C_3D_6 solid solutions. J. Chem. Phys. *58*: 4236
61. Brecher C, Krikorian E, Blane J, Halford RS (1961) Motion of molecules in condensed systems. X. The infrared spectra and structure of a single crystal of cyclopropane. J. Chem. Phys. *35*: 1097
62. Hoch MJR, Rushworth FA NMR in solid alicyclic hydrocarbons; in "Nuclear Magnetic Resonance in Chemistry." B. Pesce, ed. Academic Press, New York, NY, p. 343, 1965
63. Ruehrwein RA, Powell TM (1946) The heat capacity, vapor pressure, heats of fusion and vaporization of cyclopropane. Entropy and density of the gas. J. Am. Chem. Soc. *68*: 1063
64. Rathjens Jr GW, Freeman NK, Gwinn WD, Pitzer KS (1953) Infrared absorption spectra, structure and thermodynamic properties of cyclobutane. J. Am. Chem. Soc., *75*: 5634
65. Dunitz JD, Schomaker V (1952) The molecular structure of cyclobutane. J. Chem. Phys. *20*: 1703
66. Carter GF, Templeton DH (1953) The crystal structure of high temperature cyclobutane. Acta Cryst. *6*: 805
67. Rathjens Jr GW, Gwinn WD (1953) Heat capacity of cyclobutane. J. Am. Chem. Soc. *75*: 5629
68. Hoch MJR, Rushworth FA (1964) A nuclear magnetic resonance study of solid cyclobutane. Proc Phys. Soc. *83*: 949
69. Lord RC, Stoicheff BP (1962) High resolution Raman Spectroscopy of gases. XV. Rotational spectrum and molecular structure of cyclobutane. Canad. J. Physics, *40*: 725
70. Rushworth FA (1954) A nuclear magnetic resonance investigation of n-pentane, n-hexane and cyclopentane. Proc. Roy. Soc. (London) *A222*: 526
71. Aston G, Fink HL, Schumann SC (1943) The heat capacity and entropy, heats of transition, fusion and vaporization and the vapor pressure of cyclopentane. Evidence for a non-planar structure. J. Am. Chem. Soc. *65*: 341
72. Adams WJ, Geise HJ, Bartell LS (1970) Structure equilibrium conformation, and pseudorotation in cyclopentane. An electron diffraction study. J. Am. Chem. Soc. *92*: 5013
73. Post B, Schwartz RS, Fankuchen I (1951) X-ray investigation of crystalline cyclopentane and neohexane. J. Am. Chem. Soc. *73*: 5113
74. Andrew ER, Eades RG (1953) A nuclear magnetic resonance investigation of solid cyclohexane. Proc. Roy. Soc. (London) *A216*: 398
75. Aston JG, Szasz GJ, Fink HL (1943) The heat capacity and entropy, heats of transition, fusion and vaporization and the vapor pressure of cyclohexane. The vibrational frequencies of alicyclic ring systems. J. Am. Chem. Soc. *65*: 1135
76. Kahn R, Fourme R, Andre D, Renand M (1973) Crystal structure of cyclohexane I and II. Acta Cryst., *B29*: 131
77. Brown CW, Obremski RJ, Lippincott ER (1970) Vibrational spectra of single crystals. Polymorphic solids of cyclohexane-d_{12}. J. Chem. Phys. *52*: 2253 Obremski RJ, Brown CW, Lippincott ER (1968) Vibrational spectra of single crystals. Polymorphic solids of cyclohexane, J. Chem. Phys. *49*: 185
78. Ellett JD, Haeberlen U, Waugh JS (1970) High resolution nuclear magnetic resonance of solid perfluorohexane. J. Am. Chem. Soc. *92*: 411
79. Van Dyke Tiers G (1960) Fluorine nuclear spin resonance. Part III. The slow conformational isomerism of perfluorocyclohexane and the nature of rotational barriers in fluorocarbons. Proc. Chem. Soc. London, 389
80. Sachse H (1890) Ueber die geometrischen Isometrien der Hexamethylenderivate. Ber. *23*: 1363
81. Beckett CW, Pitzer KS, Spitzer R (1947) The thermodynamic properties and molecular structure of cyclohexane, methylcyclohexane, ethylcyclohexane, and the seven dimethylcyclohexanes. J. Am. Chem. Soc. *69*: 2488 see also *80*: 6697 (1958)
82. Pickett HM, Strauss HL (1970) Conformational structure, energy and inversion rates of cyclohexanes and some related oxanes. J. Am. Chem. Soc. *92*: 7281
83. Jensen FR, Noyce DS, Sederholm CH, Berlin AJ (1960) The energy barrier for the chair-chair interconversion of cyclohexane. J. Am. Soc., *82*: 1256
84. Anet FAL, Bourn AJR (1967) Nuclear magnetic resonance line-shape and double-resonance studies of ring inversion in cyclohexane-d_{11}. J. Am. Chem. Soc. *89*: 760

85. Finke HL, Scott DW, Gross ME, Messerly JF, Waddington G (1956) Cycloheptane, cyclooctane and 1,3,5-cycloheptatriene. Low temperature thermal properties, vapor pressure and derived chemical thermodynamic properties. J. Am. Chem. Soc. *78*: 5469
86. Anet FAL, Basus VJ (1973) Detection of a crown family conformation in cyclooctane by proton and carbon-13 nuclear magnetic resonance. J. Am. Chem. Soc. *95*: 4424 see also Anet FAL Adv. Polymer Sci., *45*: 169 (1974) and Anet FAL, Hartmann JS J. Am. Chem. Soc. *85*: 1204 (1963)
87. Bocian DF, Strauss HL (1977) Vibrational spectra, conformations and potential functions of cycloheptane and related oxepanes. J. Am. Chem. Soc. *99*: 2866
88. Pakes PW, Rounds TC, Strauss HL (1981) Vibrational spectra and potential functions of cyclooctane and some related oxocanes. J. Phys. Chem. *85*: 2476 Conformations of cyclooctane and some related oxocanes. *ibid.* 2469
89. Drotloff H "Struktur und Bewegungszustände von Cycloalkanen im kristallinen und mesomorphen Zustand. Eine Studie zum Festkörperverhalten von flexiblen Kettenmolekülen." Thesis, Albert-Ludwigs-Universität Freiburg, Fakultät Chemie, 1987
90. Anet FAL, Cheng AK, Wagner JJ (1972) Determination of conformational energy barriers in medium and large ring cycloalkanes by ^{1}H and ^{13}C nuclear magnetic resonance. J. Am. Chem. Soc. *94*: 9250
91. Drotloff H, Emeis D, Waldron RF, Möller M (1987) Chain folding and mesomorphic states of cycloalkanes. Polymer, *28*, 1200
92. Grossmann HP, Arnold R, Burkle KR (1980) IR spectroscopic investigations of cycloalkanes. Polymer Bulletin, *3*: 135
93. Fuhrer H, Gunthard HsH (1965) Schwingungsspektrum von Cyclododecan und Cyclododecan-d_2 zwischen 4000 und 50 cm^{-1}; Normalkoordinatenanalyse von zwei Cyclododecanmodellen. Helv. Chim. Acta. *48*: 236
94. Dunitz JD, Shearer HMM (1960) Die Strukturen der mittleren Ringverbindungen III. Die Struktur des Cyclodecans. Helv. Chim. Acta, *43*: 18
95. Möller M (1985) Cross-polarization-magic angle sample spinning NMR studies, with respect to the rotational isomeric states of saturated chain molecules. Adv. Polymer Sci., *66*: 59
96. Dale J (1973) Exploratory calculations of medium and large rings. Part 1. Conformational minima of cycloalkanes. Part 2. Conformational interconversions in cycloalkanes. Act. Chem. Scand., *27*: 1115, 1130
97. Kögler G "Molekulare Dynamic in festen Zustand von Cycloalkanen." Thesis, Albert-Ludwigs-Universitat, Freiburg, Fakultat Chemie, 1987
98. Arndt R (1965) NMR second moment of solid cyclododecane, J. Chem. Phys, *43*: 1885
99. Anet FAL, Rawdah TN (1978) Cyclododecane. Force-field calculations and ^{1}H NMR spectra of deuterated isotopomers. J. Am. Chem. Soc. *100*: 7166
100. Drotloff H, Rotter H, Emeis D, Moeller M (1987) The mesomorphic state of cyclotetradecane. J. Am. Chem. Soc., *109*: 7797
101. Groth P (1976) Crystal structure of cyclotetradecane at −157 °C. Acta Chem. Scand. *A30*: 155
102. Billeter E, Gunthard HsH (1958) Infrarotspektren von Cyclanen und Cyclanonen. IV. Infrarotspektren von Cyclanen. Helv. Chim. Acta, *41*: 338
103. Borgen G, Dale J (1970) Conformational analysis of 14- and 16-membered rings. Chem. Comm. 1340
104. Anet FAL, Cheng AK (1975) Conformation of cyclohexadecane. J. Am. Chem. Soc. *97*: 2420
105. Groth P (1979) The crystal conformation of cyclotetraeicosane and cyclohexaeicosane at −160 °C. Acta Chem. Scand. *A33*: 199
106. Grossmann HP (1981) Investigation of conformational transitions of cycloalkanes, especially $(CH_2)_{22}$. Polymer Bull. *5*: 137
107. Ando I, Yamanobe T, Sorita T, Komoto T, Sato H, Deguchi K, Imanari M (1984) Conformation of cyclic paraffins and polyethylene in the solid state as studied by ^{13}C CP/MAS NMR. Macromolecules *17*: 1955
108. Trzebiatowski T, Dräger M, Strobl G (1982) Die Kristallstruktur von Cycloalkanen. Makromol. Chemie, *183*: 731
109. Strobl GR, Schwickert H, Trzebiatowski T (1983) Molecular motion, defect structures and phase transitions in oligomer crystals. Ber. Bunsenges. Phys. Chem. *87*: 244

110. Lieser G, Lee K-S, Wegner G Packing of long-chain cycloalkanes in various crystalline modifications. — An electron diffraction investigation. Colloid Polymer Sci., to be published 1988
111. Groth P (1975) On the crystal structure of cyclotetradecanone at —157 °C. Acta Chem. Scand. *A29*: 374
112. Anet FAL, Cheng AK, Krane J (1973) Conformations and energy barriers in medium and large ring ketones. Evidence from ^{13}C and ^{1}H nuclear magnetic resonance. J. Am. Chem. Soc. *95*: 7877
113. Huffman HM, Eaton M, Oliver CD (1948) The heat capacities, heats of transition, heats of fusion and entropies of cyclopentene and cyclohexene. J. Am. Chem. Soc. *70*: 2911
114. Lawrenson IJ, Rushworth FA (1958) Nuclear magnetic resonance in solid cyclopentene. Proc. Phys. Soc. *72*: 791
115. Andrew ER (1963) Nuclear magnetic resonance investigation of solids. Ber Bunsen Ges. Phys. Chem. *67*: 295 For original data see El Saffar ZM, Thesis PhD Univ. College N. Wales, Bangor, 1960; also Eades RG, Finch ND, El Saffar ZM Arch. Sci (Geneva) *12*: 109 (1959) and 8th Colloque Amperé (London) 1959 (P. 109)
116. Haida O, Suga H, Seki S (1973) Three types of glassy state for cyclohexene as a single compound. Chem. Lett. 79
117. Adachi K, Suga H, Seki S (1968) Phase changes in crystalline and glassy-crystalline cyclohexanol. Bull. Chem. Soc., Japan, *41*: 1073
118. Eguchi T, Sod G, Chihara H (1975) Molecular motion and polymorphic forms in cyclohexanol as studied by nuclear magnetic resonance. J. Magn. Res. *23*: 55 for dielectric data see Adachi, K. et al., Mol. Cryst. Liq. Cryst. *18*: 345 (1972)
119. Compare Figs. 2.11 and 2.13 of Ref. 32 and see Chapter 10.3 of Ref. 46
120. Roesky HW (1979) Cyclic sulfur-nitrogen compounds. Adv. Inorg. Chem. Radiochem. *22*: 239
121. Krishnamurthy SS, Sau AC, Woods M (1978) Cyclophosphazenes, Adv. Inorg. Chem. Radiochem. *21*: 41
122. Donohue J "The Structure of the Elements." Krieger RE Publ. Co. Malabar, FL., 1974/82
123. Meyer B (1976) The structure of elemental sulfur. Adv. Inorg. Chem. Radiochem. *18*: 287
124. Warren BE, Burwell JT (1935) The structure of rhombic sulfur. J. Chem. Phys. *3*: 6
125. Sands DE (1965) The crystal structure of monoclinic (β) sulfur. J. Am. Chem. Soc. *87*: 1395
126. Templeton LK, Templeton DH (1976) Crystal structure of monoclinic sulfur. Inorg. Chem. *15*: 1999
127. Montgomery RL (1974) Monoclinic sulfur: heat capacity anomaly at 198 K caused by disordering of the crystal structure. Science, *184*: 562
128. Carberry E, West R, Glass GE (1969) Cyclic polysilanes. IV. Anion radicals and spectroscopic properties of the permethylcyclopolysilanes. J. Am. Chem. Soc., *91*: 5446
129. Carrell HL, Donohue J (1972) Crystal and molecular structure of dodecamethylcyclohexasilane. Acta Cryst. *B28*: 1566
130. Larson DW, Soltz BA, Stary FE, West R Molecular motion in brittle and plastic dodecamethylcyclohexasilane and decamethylcyclopentasilane. J. Phys. Chem. *84*: 1340; see also J. Chem. Soc. Chem. Comm. 1093 (1978)
131. Cantow H-J, Emeis D, Gronski W, Hasenhindl A, Lansberg D, Möller M, Shahab Y (1984) Macromolecular conformation studies by variable temperature ^{13}C and ^{29}Si-NMR magic angle spinning. Makromol. Chemie, Suppl. *7*, 63 see also Emeis D, Cantow H-J, Möller M (1984) A comparative study of the conformational mobility of octamethyltetrasiloxane and cyclododecane in the solid state, Polymer Bulletin, *12*: 557
132. Keyes PH, Daniels WB (1975) Octaphenylcyclotetrasilane (OPCTS): A new high temperature plastic crystal. J. Chem. Phys. *62*: 2000
133. Smith GW (1986) A calorimetric measurement of phase transformation kinetics in solid octaphenylcyclotetrasiloxane: Evidence for nucleation as the rate determining process. Mol. Cryst. Liq. Cryst. *132*: 385
134. Hikosaka M, Minomura S, Seto T (1980) Melting and solid-solid transition of polyethylene under pressure. Japan. J. appl. Phys. *19*: 1763
135. Yamamoto T, Miyaji H, Asai K (1977) Structure and properties of high pressure phase of polyethylene. Japan J. Appl. Phys. *16*: 1891

136. Bunn CW (1939) The crystal structure of long chain normal paraffin hydrocarbons. The "shape" of the CH_2-group. Trans. Farad. Soc. *35*: 483 see also Zugenmaier R, Cantow H-J Kolloid Z Z Polymere *230*: 229 (1969)
137. Bassett DC, Turner B (1972) Chain extended crystallization of polyethylene: Evidence of a new high-pressure phase. Nature (London) Phys. Sci., *240*: 146 Bassett DC, Block S, Piermarini G (1974) A new high-pressure phase of polyethylene and chain-extended growth. J. Appl. Phys. *45*: 4146
138. Yasuniva M, Nakafuku C, Takemura T (1974) Melting and crystallization process of polyethylene under high pressure. Polymer J. *4*: 526
139. Ide T, Taki S, Takemura T (1977) The high pressure and high temperature dilatometer. Japan, J. Appl. Phys. *16*: 647
140. Yasuniva M, Enoshita T, Takemura T (1976) X-ray studies of polyethylene under high pressure. Japan, J. Appl. Phys. *15*: 1421
141. Tanaka H, Takemura T (1980) Studies on the high-pressure phases of polyethylene and polytetrafluoroethylene by Raman spectroscopy. Polymer J. *12*: 355
142. Pennings AJ, Zwijnenburg A (1979) Longitudinal growth of polymer crystals from flowing solutions. VI. Melting behavior of continuous fibrillar polyethylene crystals. J. Polymer Sci., Polymer Phys. Ed. *17*: 1011
143. Maeda Y, Kanetsuna H (1975) Crystallization and melting of polyethylene under high pressure. II. Effect of pressure on melting behavior of various types of crystals. J. Polymer Sci., Polymer Phys. Ed., *13*: 637
144. Wunderlich B, Arakawa T (1964) Polyethylene crystallized from the melt under elevated pressure. J. Polymer Sci., Part A, *2*: 3697
145. Geil PH, Anderson FR, Wunderlich B, Arakawa T (1964) Morphology of polyethylene crystallized from the melt under pressure. J. Polymer Sci., Part A, *2*: 3703
146. Arakawa T, Wunderlich B (1967) Thermodynamic properties of extended chain polymethylene single crystals. J. Polymer Sci., Part C, *16*: 653
147. Wunderlich B, Melillo L (1968) Morphology and growth of extended chain crystals of polyethylene. Makromol. Chemie, *118*: 250
148. Jackson JF, Hsu TS, Brasch JW (1972) Direct microscopic observations of the crystallization process of polyethylene at high pressure, acicular bladed crystals. J. Polymer Sci., Polymer Letters *10*: 207 see also Nature (London) Phys. Sci. *239*: 45 (1972)
149. Yasuniva M, Takemura T (1974) Microscopic observation of the crystallization process under high pressure. Polymer, *15*: 661
150. Yamamoto T (1979) Nature of disorder in the high pressure phase of polyethylene. J. Macromol. Sci. — Phys. *B16*: 487
151. Nagata K, Tagashiva K, Taki S, Takemura T (1981) Ultrasonic study of high pressure phase in polyethylene. Japan. J. Appl. Phys. *19*: 985
152. Miyashita S, Asahi T, Miyagi H, Asai K (1985) A single crystal of polyethylene crystallized under high pressure. Polymer, *26*: 1791
153. Czornyj G, Wunderlich B (1977) Preparationen and study of separated single crystals of extended-chain polyethylene. J. Polymer Sci., Polymer Phys. Ed. *15*: 1905
154. Schönfeld A, Wilke W (1972) Determination of crystallite size and lattice distortion in extended chain polyethylene and their change after oxidative degradation. Kolloid Z. Z. Polym. *250*: 496
155. Wunderlich B "Macromolecular Physics, Vol. II, Crystal Nucleation, Growth, Annealing." Academic Press, New York, 1976
156. Muellér A (1932) X-ray investigation of normal paraffins near their melting points. Proc. Roy. Soc. *A138*: 514
157. Broadhurst MG (1962) An analysis of the solid phase behavior of the normal paraffins. J. Research, Natl. Bur. Stand., *66A*: 241
158. Stobl GR (1977) Molecular motion, thermal expansion, and phase transitions in paraffins: A model for polymers. J. Polymer Sci., Polymer Symposium *59*: 121 see also Colloid Polymer Sci., *254*: 170 (1976)
159. Chapman D (1982) The organization and dynamics of biomembranes. Adv. Liq. Cryst. *5*: 1
160. Takamizawa K, Ogawa Y, Oyama T (1982) Thermal behavior of n-alkanes from n-$C_{32}H_{66}$ to n-$C_{80}H_{162}$, synthesized with attention paid to high purity. Polymer J. *14*: 441

161. Even B, Strobl GR, Richter D (1980) Phase transitions of chain molecules. Disc. Farad. Soc., *69*: 19
162. Doucet J, Denicoló I, Craievich AF, Germain C (1984) X-ray study of the rotor phase of paraffins (IV): $C_{27}H_{56}$, $C_{28}H_{58}$, $C_{29}H_{60}$, $C_{30}H_{62}$, $C_{32}H_{66}$ and $C_{34}H_{70}$. J. Chem. Phys. *80*: 1647
163. Bassett DC in "Development in Crystalline Polymers", Bassett DC ed. Vol. 1, Chapter 2, Applied Science, London, 1982
164. Yamamoto T, Miyaji H, Asai K (1977) Structure and properties of high pressure phase of polyethylene, Jpn. J. Appl. Phys. *16*: 1891
165. Ungar G (1986) From plastic-crystal paraffins to liquid-crystal polyethylene: Continuity of the mesophase in hydrocarbons. Macromolecules, *19*: 1317
166. Asahi T (1984) The hexagonal phase and melt of low-molecular weight polyethylene. J. Polymer Sci., Polymer Phys. Ed. *22*: 175
167. Ungar G, Keller A (1980) Effect of radiation on the crystals of polyethylene and paraffins: 1. Formation of the hexagonal lattice and the destruction of crystallinity in polyethylene. Polymer, *21*: 1273
168. Pechhold W (1968) Molekülbewegung in Polymeren. I. Teil. Konzept einer Festkörperphysik makromolekularer Stoffe. Kolloid Z.Z. Polymere *228*, 1 See also Ref. 20
169. Dorset DL, Moss B, Wittmann JC, Lotz B (1984) The pre-melt phase of n-alkanes: Crystallographic evidence for a kinked chain structure. Proc. Natl. Acad. Sci. USA, *81*: 1913
170. Ewen B, Richter D (1978) Study of molecular motion in modification C of the n-alkane n-$C_{33}H_{68}$ by means of incoherent quasielastic neutron scattering. J. Chem. Phys. *69*: 2954
171. Maroncelli M, Strauss HL, Snyder RG (1985) The distribution of conformational disorder in the high-temperature phases of the crystalline n-alkanes. J. Chem. Phys. *82*: 2811
172. Dill KA, Flory PJ (1980) Interphases of chain molecules: Monolayers and lipid bilayer membranes. Proc. Natl. Acad. Sci., USA, *77*: 3115 and: Molecular organization in micelles and vesicles. Proc. Natl. Acad. Sci. USA, 78: 676 (1981)
173. Dehl RE (1973) NMR second moment of rotor-phase polycrystalline n-$C_{19}H_{40}$. J. Chem. Phys. *60*: 339
174. Olf HG, Peterlin A (1970) NMR study of molecular motion in oriented long chain alkanes. III. Oriented n-$C_{32}H_{66}$. J. Polymer Sci., Part A-2, *8*: 791
175. Möller M, Cantow H-J, Drottloff H, Emeis D, Lee K-S, Wegner G (1986) Phase transitions and defect structures in the lamellar surface of polyethylene and n-alkane crystallites. Magic angle spinning ^{13}C NMR studies. Makromol. Chem. *187*: 1237
176. Starkweather Jr HW, Zoller P, Jones GA, Vega AJ (1982) The heat of fusion of polytetrafluoroethylene. J. Polymer Sci., Polymer Phys. Ed., *20*: 751
177. Lau S-F, Suzuki H, Wunderlich B (1984) The thermodynamic properties of polytetrafluoroethylene. J. Polymer Sci., Polymer Phys. Ed., *22*: 379
178. Marx P, Dole M (1955) Specific heat of synthetic high polymers. V. A study of the order-disorder transition in polytetrafluoroethylene. J. Am. Chem. Soc. *17*: 4771
179. Starkweather Jr HW (1979) A comparison of the rheological properties of polytetrafluoroethylene below its melting point with certain low-molecular weight smectic states. J. Polymer Sci., Polymer Phys. Ed. *17*: 73
180. Clark ES, Muus LT (1962) Partial disordering and crystal transitions in polytetrafluoroethylene. Z. Krist *117*: 119 see also E.S. Clark, J. Makromol Phys. *B1*: 795 (1967)
181. Corradini P, Guerra G (1977) On the chain conformation of poly(tetrafluoroethylene) in the crystalline modification above 30 °C. Macromolecules, *10*: 1410
182. D'Ilario L, Giglio E (1974) Calculation of the van der Waals potential energy for polyethylene and polytetrafluoroethylene as two and three-atom chains: Rotational freedom in the crystals. Acta Cryst. *B30*: 372
183. McCall DW, Douglass DC, Falcone DR (1967) Nuclear magnetic relaxation in polytetrafluoroethylene. J. Phys. Chem. *71*: 998
184. Vega AJ, English AD (1980) Multiple-pulse nuclear magnetic resonance of solid polymers. Polymer motion in crystalline and amorphous poly(tetrafluoroethylene). Macromolecules, *13*: 1635
185. Fleming WW, Fyfe CA, Kendrick RD, Lyerla JR, Vanni H, Yannoni CS Variable temperature

high-resolution in "Polymer Characterization by ESR and NMR" ACS Symposium Series 142, Eds. Woodward AE, Bovey FA, Washington DC 1980; see also Macromolecules, *13*: 460 (1980)
186. Natarajan RT, Davidson T (1972) Kinetics of the 20 °C phase transition in polytetrafluoroethylene. J. Polymer Sci., Polymer Phys. Ed. *10*: 2209
187. Bates TW, Stockmayer WH (1968) Conformational energies of perfluoroalkanes. III. Properties of polytetrafluoroethylene. Macromolecules *1*: 17
188. Tanaka H, Takemura T (1980) Studies on the high-pressure phases of polyethylene and polytetrafluorethylene by Raman spectroscopy. Polymer J., *12*: 355
189. Matsushige K, Enoshita R, Ide T, Yamauchi N, Taki S, Takemura T (1977) Fine structure of the III–I transition and molecular motion in polytetrafluoroethylene. Jpn. J. Appl. Phys. *16*: 681
190. Melillo L, Wunderlich B (1972) Extended chain crystals VIII. Morphology of polytetrafluoroethylene. Kolloid Z. Z. Polymere *250*: 417
191. Bassett D, Davitt R (1974) On the crystallization phenomena in polytetrafluoroethylene. Polymer *15*: 721
192. Campos-Vallette M, Ray-Lafou M, Lagnier R (1982) Vibrational and calorimentric study of phase transitions in n-C_6F_{14} and n-C_8F_{18} crystals. Chem. Phys. Lett. *89*: 189
193. Bastiansen O, Hadler E (1952) Molecular structures of dodecafluoro-n-pentane, hexadecafluoro-n-heptane and dodecafluoro-isopentane. Act. Chem. Scand. *6*: 214
194. Chantry GW, Nicol EA, Jones RG, Willis HA, Cudby MEA (1977) On the vibrational assignment problem for polytetrafluoroethylene: 1. The infrared spectrum. Polymer *18*: 37
195. Bunn CW, Howells ER (1954) Structures of molecules and crystals of fluorocarbons. Nature *174*: 549
196. Smith P, Gardner KH (1985) Dissolution of poly(tetrafluoroethylene) Macromolecules, *18*: 1222
197. Hasegawa R, Kobayashi M, Tadokoro H (1972) Molecular conformation and packing of poly(vinylidene fluoride). Polymer J. *3*: 391 Hasegawa R, Takahashi Y, Chatani Y, Tadokoro H (1972) Crystal structures of three crystalline forms of poly(vinylidene fluoride). Polymer J. *3*: 600
198. Takahashi Y, Tadokoro H, Odajima A (1980) Kink bands in form I of poly(vinylidene fluoride). Macromolecules *13*: 1318
199. Matsushige K, Takemura T (1980) Crystallization of macromolecules under high pressure. J. Crystal Growth *48*: 343
200. Gal'perin YeL, Strogalin YuV (1965) The symmetry and dimension of the unit cell of polytrifluoroethylene. Vysokomol. Soyed. *7*: 16 Engl. translat. Polymer Sci. USSR, *7*: 15 (1965)
201. Lovinger A, Cais RE (1984) Structure and morphology of polytrifluoroethylene. Macromolecules, *17*: 1939
202. Loufakis K, Wunderlich B (1986) Heat capacities of linear macromolecules containing chlorine and fluorine. Polymer, *27*: 563
203. Loufakis K, Wunderlich B (1987) Thermal analysis of the conformational disorder in semicrystalline poly(vinylidene fluoride) and polytrifluoroethylene. Macromolecules, *20*: 2474
204. Miyamoto Y, Nakafuku C, Takemura T (1972) Crystallization of polychlorotrifluoroethylene. Polymer J. *3*: 120
205. Takahashi Y, Matsubara Y, Tadokoro H (1981) Crystal structure of form II of poly(vinylidene fluoride). Macromolecules *16*: 1588
206. Lovinger A (1981) Unit cell of the γ-phase of poly(vinylidene fluoride). Macromolecules, *14*: 322 also Takahashi Y, Tadokoro H Macromolecules, *13*: 1317 (1980)
207. Loufakis K, Miller KJ, Wunderlich B (1986) On the conformational isomerism of poly(vinylidene fluoride). Macromolecules, *19*: 1217
208. Miyamoto Y, Miyaji H, Asai K (1980) Anisotropy of dielectric relaxation in crystal form II of poly(vinylidene fluoride). J. Polymer Sci., Polymer Phys. Ed. *18*: 597
209. Takahashi Y, Miyaji K (1983) Long-range order parameter of form II of poly(vinylidene fluoride) and molecular motion in the α_c relaxation. Macromolecules, *16*: 1789
210. Takahashi Y, Tadokoro H (1984) Structure and disorder of poly(vinylidene fluoride). Ferroelectrics *57*: 187

211. Takahashi Y (1983) Packing disorder in form II of poly(vinylidene fluoride); influence of elongation and annealing temperature. Polymer J. *15*: 733
212. Takahashi Y, Matsubara Y, Tadokoro H (1982) Mechanisms for crystal phase transformations by heat treatment and molecular motion in poly(vinylidene fluoride). Macromolecules *15*: 334
213. Prest J. Luca J (1978) The formation of the γ-phase from the α and β polymorphs of poly(vinylidene fluoride). J. Appl. Phys. *49*, 5042; see also *46*: 4136 (1975)
214. Lovinger A (1980) Crystalline transformations in spherulites of poly(vinylidene fluoride). Polymer *21*: 1317
215. Lovinger AJ, Davis DD, Cais RE, Kometani JM (1986) On the Curie temperature of poly(vinylidene fluoride). Macromolecules, *19*: 1491
216. Tashiro K, Takano K, Kobayashi M, Chatani Y, Tadokoro H (1981) Structural study on ferroelectric phase transition of vinylidene fluoride-trifluoroethylene random copolymers. Polymer, *22*: 1312 *25*: 195 (1984)
217. Lovinger AJ, Furukawa T, Davis GT, Broadhurst MG (1983) Crystallographic changes characterizing the Curie transition in three ferroelectric copolymers of vinylidene fluoride and trifluoroethylene. Polymer, *24*: 1225, 1233
218. Green JS, Rabe JP, Rabolt JF (1986) Studies of chain-conformation above the Curie point in a vinylidene fluoride/trifluoroethylene random copolymer. Macromolecules, *19*: 1725
219. Wu ZQ, Dann V, Cheng SZD, Wunderlich B Fast DSC applied to the crystallization of polypropylene. J. Thermal Analysis, to be published, 1988
220. Natta G, Peraldo M, Corradini P (1959) Modificazione mesomorpha smettica del polipropilene isotattico. Rend. Accad. Naz. Lincei, *24*: 14
221. Zannetti R, Celotti GC, Fichera A, Francesconi R (1969) The structural effects of annealing time and temperature on the paracrystal-crystal transition in isotactic polypropylene. Makromol. Chemie *128*: 137
222. Miller RL (1960) On the existence of near-range order in isotactic polypropylenes. Polymer *1*: 135
223. Corradini P "The Stereochemistry of Macromolecules". Dekker, New York, N.Y. Vol. 3, 1968
224. Bunn A, Cudby MEA, Harris RK, Parker KJ, Say BJ (1982) High resolution ^{13}C n.m.r. spectra of solid, isotactic polypropylene. Polymer *23*: 694
225. Grebowicz J, Lau S-F, Wunderlich B (1984) The thermal properties of polypropylene. J. Polymer Sci., Polymer Symposium *71*: 19
226. Petraccone V, DeRose C, Guerra G, Tuzi A (1984) On the double peak shape of the endotherms of isothermally crystallized isotactic polypropylene samples. Makromol. Chem., Rapid Commun., *5*: 631
227. DeRose C, Guerra G, Napolitano R, Petraccone V, Pirozzi B (1984) Conditions for the α (1) to α (2) transition in isotactic polypropylene samples. Europ. Polym. J. *20*: 937
228. Carfagna C, DeRose C, Guerra G, Petraccone V (1984) Recrystallization kinetics of isotactic polypropylene. Polymer, *25*: 1462
229. Nakafuku C (1981) High pressure d.t.a. study on the melting and crystallization of isotactic polypropylene. Polymer *22*: 1673
230. Lovinger AJ, Chua JO, Gryte CC (1977) Studies on the α and β forms of isotactic polypropylene by crystallization in a gradient. J. Polymer Sci., Polymer Phys. Ed. *15*: 641
231. Geil PH, Hsu CC (1984) Crystallization of polypropylene and poly(vinylidene fluoride) from the glassy state. J. Phys. Soc., 412
232. Poddubny VI, Lavrentyev UK, Shevelev VA, Mikhailova NV, Sidorovich AV (1984) Melt crystallization of polypropylene: Structure formations. Polymer Bulletin, *11*: 41
233. Hendra PJ, Vile J, Willis HA, Zichy V (1984) The effect of cooling rate upon the morphology of quenched melts of isotactic polypropylenes. Polymer, *25*: 785
234. Suehiro K, Takayanagi M (1970) Structural studies of the high temperature form of trans-1,4-polybutadiene crystal. J. Macromol. Sci., Phys. *B4*: 39
235. Natta G, Corradini P (1959) Conformation of linear chains and their mode of packing in the crystal state. J. Polymer Sci. *39*: 29 see also Natta G, Corradini P, Porri D Rend. Acad. Nazl. Lincei *20*: 728 (1956)
236. Möller M (1987) Structure and dynamics of the high temperature polymorphs of trans-1,4-polybutadiene. Polymer Preprints, Am. Chem. Soc. *28* (2): 395

237. Iwayanagi S, Miura J (1965) Nuclear magnetic resonance study of solid phase transition of trans-1,4-polybutadiene. Rept. Progr. Polymer Phys. Japan *8*: 303
238. Finter J, Wegner G (1981) The relation between phase transition and crystallization behavior of 1,4-trans-poly(butadiene). Makromol. Chemie *182*: 1859
239. Iwayanagi S, Sakurai I, Sakurai T (1968) X-ray structure analysis of trans-1,4-polybutadiene. Macromol. Sci. Phys. *B2*: 163
240. Corradini P (1969) On the chain conformation of the high temperature polymorphs of trans-1,4-polybutadiene. Polymer Letters, *7*: 211 see also J. Polymer Sci., Symposia *50*: 327 (1975)
241. Bautz G, Leute U, Dollhopf W, Haegele PC (1981) On the solid state phase of poly(trans-1,4-butadiene). Colloid and Polymer Sci. *259*: 714
242. Personal communication Dr. Eisele, Bayer AG, Leverkusen
243. Dreyfuss P, Keller A (1970) A simple chain refolding scheme for the annealing behavior of polymer crystals. J. Polymer Sci., Part B. *8*: 253
244. Grebowicz J, Cheng SZD, Wunderlich B (1986) Kinetics of transitions involving condis crystals. J. Polymer Sci., Part B: Polymer Physics, *24*: 675
245. Tadokoro H "Structure of Crystalline Polymers." Wiley-Interscience, New York, N.Y. 1979; for trans-1,4-poly(2-methylbutadiene) see also Takahashi Y, Sato T, Tadokoro H, Tanaka Y (1973) J. Polymer Sci., Polymer Phys. Ed., *11*: 233
246. Natta G, Corradini P (1960) The crystal structure of cis-1,4-polybutadiene. Nuovo Cimento, Suppl. *15*: 111 see also Angew. Chemie *68*: 615 (1956) and Nyburg SC (1954) Acta Cryst. *7*: 385
247. Edwards BC, Phillips PJ (1975) The structure of the high pressure phase of cis-polyisoprene, J Mat Sci., *10*: 1233 see also J. Polymer Sci., *B10*: 321 (1972) and Polymer *15*: 491 (1974)
248. Phillips PJ, Edwards BC (1976) High pressure phases in polymers. III. The nature of the high pressure phase in cis-polyisoprene. J. Polym. Sci. Polymer Phys. Ed. *14*: 377 see also *ibid*, *13*: 1819: 2117 (1975) and *14*: 391 (1976)
249. Statton WO (1959) Directional "crystallization" of polymers. N.Y. Acad. Sci., *83*: 27 see also the appendix in Kast W (1955) Angew. Chem. *67*: 592
250. Gogolewski S, Pennings AJ (1973) Crystallization of polyamides under elevated pressure. Nylon 6. Polymer *14*: 463 see also *18*: 647, 654 (1977)
251. Gogolewski S, Pennings AJ (1977) Crystallization of polyamides under elevated pressure: 5. Pressure-induced crystallization from the melt and annealing of folded chain crystals of nylon 11, poly(aminoundecaneamide) under pressure. Polymer *18*: 660
252. Stamhuis JE, Pennings AJ (1977) Crystallization of polyamides under elevated pressure: 6. Pressure-induced crystallization from the melt and annealing of folded chain crystals of nylon 12, polyaurolactam under pressure. Polymer *18*: 667
253. Hiramatsu N, Hirakawa S (1982) Melting and transformation behavior of gamma form Nylon 6 under high pressure. Polymer J. *14*: 165
254. Tadokoro H, Kobayashi M, Mori K, Takahashi Y, Taniyama S (1969) Structural studies of polyallene. J. Polymer Sci., Part C, *22*: 1031
255. Petraccone V, Ganis P, Corradini P, Montagnoli G (1972) A case of conformational isomorphism in polymers. The crystal structure of isotactic poly-(S)-3-methylpentene-1, European Polymer J. *8*: 99
256. Gaur U, Lau S-F, Wunderlich B (1983) Heat capacities and other thermodynamic properties of linear macromolecules IX. Aromatic and Inorganic polymers. J. Phys. Chem. Ref. Data, *12*: 91
257. Wesson J, Wunderlich B New measurements in our laboratory, to be published in 1988/89
258. Wiedemann HG, Wunderlich B, Wesson JP (1988) Mesophase transitions of polydiethylsiloxane. Mol. Crys. Liq. Cryst, *155*, 469
259. Papkov VS, Godovsky Yu, Svistunov VS, Litvinov VM, Zhdanov AA (1984) DSC studies of phase transitions in poly(diethyl siloxane). J. Polymer Sci., Polymer Phys. Ed., *22*: 3617
260. Beatty CL, Pochan JM, Froix MF, Hinman DD (1975) Liquid crystalline type order in polydiethylsiloxane. Macromolecules *8*: 547
261. Tsvankin DYa, Papkov VS, Zhukov VP, Godovsky YuK, Svistunov VS, Zhdanov AA (1985) Structure and phase transitions in poly(diethylsiloxane). J. Polymer Sci., Polymer Phys. Ed., *23*: 1043

262. Grebowicz J, Wesson J, Loufakis K, Miller K, Wunderlich B On the conformational isomerism of poly(diethyl siloxane). Macromolecules, to be published 1988
263. Froix MF, Beatty CL, Pochan JM, Hinman DD (1975) Nuclear spin relaxation in poly(diethylsiloxane). J. Polymer Sci., Polymer Phys. Ed. *13*: 1269
264. Godovsky YuK, Makarova NN, Papkov VS, Kuzmin NN (1985) Crystalline phases and a mesophase in poly(dipropylsiloxane). Makromol. Chemie, Rapid Comm. *6*: 443 Influence of molecular weight on the thermotropic mesophase behavior of polypropylsiloxane. ibid, *6*: 797 (1985)
265. Raboldt JF, Hofer D, Miller RD, Fickes GN (1986) Studies of chain conformational kinetics in poly(di-n-alkylsilanes) by spectroscopic methods. Macromolecules, *19*: 611
266. Schneider NS, Desper CR, Beres JJ Mesomorphic structures in polyphosphazenes. Blumstein A (ed) "Liquid Crystalline order in Polymers." Academic Press, New York NY 1978, p. 299–325
267. Singler RE, Schneider NS, Hagnamer GL (1975) Polyphosphazenes: synthesis-properties-applications. Polymer Eng. Sci. *15*: 321
268. Schneider NS, Desper CR, Singler RE (1976) The thermal transitions behavior of polyorganophosphazenes. J. Appl. Polymer Sci., *20*: 3087
269. Masuko T, Simeone RL, Magill JH, Plazek DJ (1984) Physical properties of polyphosphazenes. 2. Poly[bis(trifluoroethoxy)phosphazene]. Macromolecules *17*: 2857
270. Alexander MN, Desper CR, Sagalyn PL, Schneider NS (1977) Nuclear magnetic resonance and X-ray study of the mesomorphic transition in poly[bis(2,2,2-trifluoroethoxy)-phosphazene]. Macromolecules *10*: 721
271. Kojima M, Magill JH (1985) Spherulitic crystallization in poly(bis(trifluoroethoxy)phosphazene), PBFP. Polymer, *26*: 1971 Morphology of polyphosphazene crystals. Polymer Comm. *24*: 329 (1983)
272. Kojima M, Magill JH (1986) Fracture surface morphology of poly[bis(trifluoroethoxy)phosphazene]. J. Mat. Sci., *21*: 2651 see also Makromol. Chemie *186*: 649 (1985)
273. Magill JH, Petermann J, Rieck U (1986) Morphology of polyphosphazenes: oriented and unoriented. Colloid and Polymer Sci., *264*: 570
274. Kojima M, Satake H, Masuko T, Magill JH Crystal modifications of poly[bis(phenoxy)phosphazene]. PBPP. J. Materials Sci., to be published, 1988
275. Magill JJ, Riekel C (1986) A phase transition study in poly[bis(p-methylphenoxy)phosphazene] using synchrotron radiation. Makromol. Chemie, Rapid Comm. *7*: 287
276. Samulski ET (1985) Macromolecular structure and liquid crystallinity. Farad. Discuss. Chem. Soc. *79*: 7
277. Huffman HM, Parks GS, Daniels AC (1930) Thermal data on organic compounds. VII. The heat capacities, entropies and free energies of twelve aromatic hydrocarbons. J. Am. Chem. Soc. *52*: 1547
278. Andrews ER, Eades RG (1953) A nuclear magnetic resonance investigation of three solid benzenes, Proc. Roy. Soc., London, *A218*: 537
279. Haeberlen U, Maier G (1967) Protonenspinrelaxation in kristallinem Benzol. Z. Naturforsch. *22a*: 1236
280. Bacon GE, Curry NA, Wilson SA (1964) A crystallographic study of solid benzene by neutron diffraction. Proc. Roy. Soc., London, *A279*: 98
281. Cox EG, Cruickshank DW, Smith JAS (1958) The crystal structure of benzene at −3 °C. Proc. Roy. Soc., London, *A247*: 1
282. Van Steenwinkel R (1969) The spin lattice relaxation of the nuclear dipolar energy in some organic crystals with slow molecular motions. Z. Naturforsch. *24a*: 1526
283. Lauer O, Stehlik D, Hausser KH (1972) Nuclear Zeeman and dipolar relaxation due to slow motion in aromatic single crystals. J. Magn. Resonances *6*: 524
284. Pawley GS (1985) The molecular dynamics of a napthalene crystal. Solid State Comm. *53*: 817
285. Fourme R (1972) Étude par diffraction X des structures cristallines du furanne à la pression atmosphérique. Acta Cryst. *B28*: 2984
286. Guthrie GB Jr, Scott DW, Hubbard WN, Katz C, McCullough JP, Gross ME, Williamson KD, Waddington G (1952) Thermodynamic properties of furan. J. Am. Chem. Soc., *74*: 4662
287. Fried F (1966) Étude résonance magnétique nucléaire large bande du furanne cristallisé. Comp. rend. 262C: 1497

288. Andon RJL, Counsell JF, Hales JL, Lees EB, Martin JF (1968) Thermodynamic properties of fluorine compounds. Part VII. Heat capacity and entropy of penta fluoro chlorobenzene and pentafluorophenol. J. Chem. Soc. A: 2357
289. Brot C, Darmon I (1970) Orientational freedom of molecules in crystals. II. High-order transition by progressive decorrelation of orientations. A Monte Carlo calculation. J. Chem. Phys. *53*: 2271 for NMR data see also: J. chim. Phys. *64*: 1061 (1967) and, Eveno M and Meinnel J (1966) J. chim. Phys. *63*: 108
290. Frankosky M, Astron JG (1965) The heat capacity and entropy of hexamethylbenzene from 13 to 340 K. An estimate of the internal rotation barrier. J. Phys. Chem. *69*: 3126
291. Allen PS, Cowking A (1967) Nuclear magnetic resonance study of molecular motions in hexamethylbenzene. J. Chem. Phys. *47*: 4286
292. Bastiansen O (1949) The molecular structure of biphenyl and some of its derivatives. Act. Chem. Scand. *3*: 408 See also Bastiansen O, Traetteberg M Tetrahedron *17*: 147 (1962)
293. d'Annibale A, Lunazzi L, Boicelli AC, Macciantelli D (1973) The conformational problem of biphenyl in solution as investigated by the liquid crystal magnetic resonance spectrum of 3,3',5,5'-tetrachlorobiphenyl. J. Chem. Soc. Perkins II: 1396 see also Schmid ED, Brosa BJ (1972) Chem. Phys., *56*: 6267
294. Trotter J (1961) The crystal and molecular structure of biphenyl. Acta Cryst. *14*: 1135
295. Hargreaves A, Hasan-Rizvi S (1962) The crystal and molecular structure of biphenyl. Acta Cryst. *15*: 365
296. Dworkin A, Cailleau H (1980) Heat capacity of biphenyl and its phase transitions at low temperature. J. Physique, Lett. *41*: L-83
297. Charbonneau G-P, Delugeard Y (1977) Biphenyl: Three-dimensional data and new refinement at 293 K. Acta Cryst. *B33*: 1586
298. Charbonneau G-P, Delugeard Y (1976) Structural Transitions in polyphenyls. III. Crystal structure of biphenyl at 110 K. Acta Cryst. *B32*: 1420
299. Cailleau H, Baudour JL, Zeyen CME (1979) Structural phase transition in polyphenyls. VIII. A neutron diffraction study of the structural phase transition in biphenyl d_{10}. Acta Cryst. *B35*: 426
300. Friedman PS, Kopelman R, Prasad PN (1974) Spectroscopic evidence for a continuous change in molecular and crystal structure: Deformation of biphenyl in the low temperature solid. Chem. Phys. Lett. *24*: 15
301. Cailleau H, Baudour J-L, Meinnel J, Dworkin A, Moussa F, Zeyen CME (1980) Double-well potentials and structural phase transitions in polyphenyls. Farad. Disc. Chem. Soc. *69*: 7
302. Cailleau H, Dworkin A (1979) Calorimetric study of the phase transition of p-terphenyl. Mol. Cryst. Liq. Cryst., *50*: 217
303. Toudic B, Lechner RE (1984) Observation of the effects of critical phenomena in paraterphenyl on quasielastic incoherent neutron spectra: II. J. Phys. C., *17*: 5503 see also Toudic B, Cailleau H, Lechner RE, Petry W (1986) Phys. Rev. Letters, *56*: 347
304. Baudour JL, Delugeard Y, Cailleau H (1976) Transition structurale dans les polyphenyles. I. Structure crystalline de la phase basse temperature du p-terphenyle a 113 K. Acta Cryst. *B32*: 150
305. Rietveld HM, Maslen EN, Clews CJB (1970) An X-ray and neutron diffraction refinement of the structure of p-terphenyl. Acta Cryst. *B26*: 693
306. Baudour JL (1972) Facteur de temperature et librations de grande amplitude. Application an p-terphenyle. Acta Cryst. *B28*: 1649
307. Baudour JL, Cailleau H (1977) Structural phase transitions in polyphenyls. V. Double-well potential in the disordered phase of p-terphenyl from neutron (200 K) and X-ray (room temperature) diffraction data. Acta Cryst. *B33*: 1773
308. Toudic B, Gallier J, Rivet P, Delugeard Y (1985) Proton magnetic resonance study of some crystalline p-polyphenyls: Evidence of relaxation by slow molecular motions above room temperature. Chem. Phys. *99*: 275
309. Wasicki J, Radomska M, Radomski R (1982) Heat capacity of diphenyl, p-terphenyl and p-quaterphenyl from 180 K to their melting points. J. Thermal Analysis *25*: 509
310. Smith GW (1979) Phase behavior of some linear polyphenyls. Mol. Cryst. Liq. Cryst. *49*: 207
311. Baudour JL, Delugeard Y, Rivet P (1978) Structrural phase transition in polyphenyls.

VI. Crystal structure of the low-temperature ordered phase of p-quaterphenyl at 110 K. Acta. Cryst. *B34*: 625
312. Delugeard Y, Deguche J, Baudour J-L (1976) Structural transitions in polyphenyls. II. The crystal structure of the high temperature phase of quaterphenyl. Acta. Cryst. *B32*: 702
313. Bottom BA, Prasad, PN (1978) Phase transitions in polyphenyls: Raman spectra of p-terphenyl and p-quaterphenyl in the solid state. Chem. Phys. *35*: 331
314. Ecolivet C, Toudic B, Sanquer M (1981) Brillouin scattering study of phase transitions in p-polyphenyls. J. Physique, *C6*: 779
315. Irvine PA, Wu DC, Flory PJ (1984) Liquid-crystalline transitions in homologous p-phenylenes and their mixtures. Part I. — Experimental results. J. Chem. Soc., Farad. Trans. I, *80*: 1795
316. Flory PJ, Ronca G (1979) Theory of systems of rodlike particles. II. Thermotropic systems with orientation-dependent interactions. Mol. Cryst. Liq. Cryst. *54*: 311
317. Rabinovich IB, Sapozhnikov VN, Sergeev VA, Karyakin NY, Krylova GP, Korshak VV (1971) Thermodynamics of poly-p-phenylene. Dokl. Acad. Nauk SSSR, *200*: 890
318. Cheng SZD, Lim S, Judovits LH, Wunderlich B (1987) Heat capacities of high melting polymers containing phenylene groups. Polymer, *28*: 10
319. Kawaguchi A, Petermann J (1986) Paracrystalline nature of poly(p-phenylene). Mol. Cryst. Liq. Cryst. *133*: 189
320. Kelker H, Hatz R "Handbook of Liquid Crystals", Verlag Chemie, Weinheim 1980
321. Bernal JD, Crowfoot D (1933) Crystalline phases of some substances studied as liquid crystals. Trans. Farad. Soc. *29*: 1032 For newer and more complete data see Krigbaum WR, Chatani Y, Barber PG (1970) Acta Cryst. *B26*: 97 Carlile CH, Smith CH (1971) Acta Cryst. *B27*: 1068
322. Bernstein J, Schmidt GMJ (1972) Conformational studies. Part IV. Crystal and molecular structures of the metastable form of N-(p-chlorobenzylidene)-p-chloroaniline, a planar anil. J. Chem. Soc. Perkin Trans. II. 951
323. Bryan RF, Hartley P, Miller RW, Shen M-S (1980) An X-ray study of the p-n-alkoxybenzoic acids. Part VI. Isotypic crystal structures of four smectogenic acids having seven, eight, nine, and 10 alkyl chain atoms. Mol. Cryst. Liq. Cryst., *62*: 281
324. Bryan RF, Hartley P, Miller RW (1980) An X-ray study of the p-n-alkoxybenzoic acids. Crystal structures of related forms p-n-hexoxy and p-n-octoxybenzoic acids. Mol. Crystl. Liq. Cryst. *62*: 311
325. Herbert AJ (1967) Transition temperatures and transition energies of the p-n-alkoxy benzoic acids, from n-propyl to n-octadecyl. Trans. Farad. Soc. *63*: 555
326. Rowell JC, Phillips WD, Panar M (1965) NMR studies of some liquid crystal systems. J. Chem. Phys. *43*: 3442
327. Pines A, Chang JJ (1974) Study of the isotropic-nematic-solid transitions in a liquid crystal by carbon-13-proton double resonance. Phys. Rev. *A10*: 946 see also Pines A, Rubin DJ, Allison S (1974) Phys. Rev. Letters, *33*: 1002
328. Schmiedel H, Hillner B, Grande S, Losche A, Limmer St (1980) Interpretation of Proton NMR Lineshape of Liquid Crystals. J. Magn. Res. *40*: 369
329. Kumagai M, Soda G, Chihara H (1981) Nuclear magnetic relaxation of nemattic liquid crystals, N-(4-methoxybenzylidene)-4-n-butylaniline and N-(2-hydroxy-4-methoxybenzylidene)-4-n-butylaniline. J. Magn. Res. *42*: 28
330. Arumugam S, Bhat SV, Kunar N, Ramanathan KV, Srinivasan R (1985) NMR and X-ray investigations of a plastic crystalline phase in MBBA. Mol. Cryst. Liq. Cryst. *126*: 161
331. Janik JA, Janik JM, Mayer J, Sciesinska E, Scienski J, Twardowski J, Waluga T, Witko W (1975) Calorimetric and infrared study of the phase situation in solid MBBA. J. Physique, *36*: Cl-159
332. Lydin JE, Kesler JO (1975) Phase transitions on warming fast quenched MBBA. J. Physique *36*: C1-153
333. LeBrumant J, Jaffrain M, Nguyen At (1979) Etude du polymorphisme de p-methoxybenzylidene-p-n-butylaniline (MBBA) per spectroscopie vibrationelle infrarouge et Raman. J. Physique, *40*: C3-193
334. Sorai M, Tsuji K, Suga H, Seki S (1980) Studies on disk-like molecules. I. Heat capacity of benzene-hexa-n-hexanoate, Mol. Cryst. Liq. Cryst. *59*: 33

335. Sorai M, Suga H (1981) Studies on mesogenic disk-like molecules. II. Heat capacity of benzene hexa-n-heptanoate from 13 to 393 K. Mol. Cryst. Liq. Cryst. *73*: 47
336. Sorai M, Yoshioka H, Suga H (1982) Studies on mesogenic disk-like molecules. III. Heat capacity of benzene-hexa-n-octanoate from 13 to 393 K. Mol. Cryst. Liq. Cryst. *84*: 39
337. van Hecke GR, Kaji K, Sorai M (1986) Heat capacity of the discotic mesogen 2,3,6,7,10,11-hexa-n-octanoyl triphenylene: A complex solid state polymorphism. Mol. Cryst. Liq. Cryst. *136*: 197
338. Coker TG, Wunderlich B, Janz GJ (1969) Melting mechanisms of ionic salts: Tetra-n-amyl-ammonium thiocyanate. Trans. Farad. Soc. *65*: 3361
339. Janz GJ "Molten Salts Handbook." Academic Press, New York, NY, 1967
340. Iwamoto K, Ohuki Y, Sawada K, Seno N (1981) Solid-solid phase transitions of long-chain n-alkyltrimethylammonium halids. Mol. Cryst. Liq. Cryst., *73*: 95
341. Busico V, Corradini P, Vacatello M (1982) Thermal behavior and observation of a smectic phase in n-pentadecylammonium chloride. J. Phys. Chem. *86*: 1033
342. Busico V, Cerniccia P, Corradini P, Vacatello M (1983) Polymorphism in anhydrous and amphilic systems. Long-chain primary n-alkylammonium chlorides. J. Phys. Chem., *87*: 1631
343. Lindau J, Diele S, Krüger H, Dörfler H-D (1981) Thermisches Phasenverhalten von Thallium (I)-salzen aliphatischer Karbonsäuren. Z. phys. Chem. (Leipzig) *262*: 775
344. Lindau J, König H-J, Dörfler H-D (1983) Systematisierung der Tieftemperaturmodifikationen homologer Thallium(I)-Salze aliphatischer Carbonsäuren durch ihre Mischbarkeitsbeziehungen. Colloid Polymer Sci. *261*: 236
345. Busico V, Ferraro A, Vacatello M (1984) Polymorphism and liquid crystalline behavior of lithium n-hexadecanoate. J. Phys. Chem., *88*: 4055
346. Skoulios AE, Luzzati V (1961) La structure des colloides d'association. III. Descritpion des phases mésomorphes des savons de sodium purs, recontrées au-dessus de 100 °C. Acta Cryst. *14*: 278 see also Gallot B, Skoulios AE, Kolloid Z Z (1966) Polym. *209*: 164
347. Förster G, Brezesinski G, Gerlach E, Mädicke A, Dörfler H-D (1981) Einfluß verschiedener Vorbehandlungen auf die Strukturbildung und thermischen Umwandlungserscheinungen ausgewählter Natrium-Seifen. Z. phys. Chem. (Leipzig), *6*: 1009
348. Chapman D, Byrne R, Shipley GG (1966) The physical properties of phospholipids. I. Solid state and mesomorphic properties of some 2,3-diacyl-D,L-phosphatidylethanolamines. Proc. Roy. Soc. *A290*: 115
349. Chapman D, Salsbury NJ (1966) Physical studies of phospholipids. Part 5. Proton magnetic resonance studies of molecular motion in some 2,3-diacyl-D,L-phosphatidyl ethanolamines. Trans. Farad. Soc. *62*: 2607
350. Busico V, Ferraro A, Vacarello M (1986) A study of the "rotor" phase of 16-hydroxy-lithium-hexadecanoate. J. Chem. Phys. *84*: 471
351. Russell TP, Rabolt JF, Twieg RJ, Siemens RL, Farmer BL (1986) Structural characterization of semifluorinated n-alkanes. 2. Solid-solid transition behavior. Macromolecules, *19*: 1135
352. Pugh C, Höpken J, Möller M (1988) Amphiphilic molecules with hydrocarbon and fluorocarbon segments. Polymer Preprints Div. Polymer Sci., Am. Chem. Soc. *29(1)*: 460
353. See for example the extensive reviews collected in "Liquid Crystalline Polymers" Gordon M, Plate NA (eds) (1984) Adv. Polymer. Sci., Volumes 59 to 61
354. Boeffel C, Spiess HW, Hisgen B, Ringsdorf H, Ohm H, Kirste RG (1986) Molecular order of spacer and main chain in polymeric side-group liquid crystals. Makromol. Chemie, Rapid Comm. *7*: 777
355. Kirkpatrick DE, Wunderlich B (1985) Thermal analysis of the phase transitions of poly(p-xylylene). Makromol. chem. *186*: 2595 see also J. Polymer Sci., Polymer Phys. Ed. *24*: 931 (1986)
356. Isoda S, Kawaguchi A, Katayama K-I (1984) High temperature phases of poly(p-xylylene). J. Polymer Sci., Polymer Phys. Ed., *22*: 669
357. Iwamoto R, Wunderlich B (1973) Crystal Structure of poly-p-xylylene. I. The α-form. J. Polymer Sci., Polymer Phys. Ed., *11*: 2403
358. Isoda S, Tsuji M, Ohara M, Kawaguchi A, Katayama K (1983) Structural analysis of β-form poly(p-xylylene) starting from a high-resolution image. Polymer *24*: 1155
359. Sikkema DJ, Hoogland P (1986) Poly(7-oxabicyclo[2.2.1]heptane-*co*-tetrahydrofurnane): Polymerization, polymer characterization and spinning. Polymer *27*: 1443

360. Siegmann A, Harget PJ (1980) Melting and crystallization of poly(ethylene terephthalate) under presure. J. Polymer Sci., Polymer Phys. Ed. *18*: 2181
361. English AD (1984) Macromolecular dynamics in solid poly(ethylene terephthalate): ^{1}H and ^{13}C solid state NMR, Macromolecules *17*: 2182
362. Jelinski LW, Dumais JJ, Watnick PI, Engel AK, Sefcik MD (1983) NMR relaxation in poly(butylene terephthalate) and poly(butylene terephthalate)-containing segmented copolymers. Macromolecules *16*: 409
363. Ober CK, Jin J-I, Lenz R (1984) Liquid crystal polymers with flexible spacers in the main chain. Advan. Polymer Sci., *59*: 103
364. Grebowicz J, Wunderlich B (1983) Phase transitions in mesophase macromolecules. IV. The transitions in poly(oxy-2,2'-dimethylazoxybenzene-4,4'-diyloxydodecanedioyl). J. Polymer Sci., Polymer Phys. Ed. *21*: 141
365. Roviello A, Sirigu A (1980) A mesomorphic polymer: poly(oxydodecanedioloxy-1,4-phenylene-2-methylvinylene-1,4-phenylene). Macromol. Chem. *181*: 1799
366. Blumstein A, Thomas O (1982) Odd-even effect in thermotropic liquid crystalline 4,4-dihydroxy-2,2'-dimethylazoxybenzene-alkanedioic acid polymers. Macromolecules *15*: 1264
367. Samulski ET, Ganthier MM, Blumstein RB, Blumstein A (1984) Alkyl chain order in a linear polymeric liquid crystal. Macromolecules *17*: 479
368. Müller K, Meier P, Kothe G (1985) Multipulse dynamic NMR of liquid crystalline polymers Progr. NMR Spectr. *17*: 211
369. Müller K, Kothe G (1985) ^{2}H NMR studies of liquid crystal main chain polymers. Ber. Bunsen Ges. Phys. Chem. *89*: 1214
370. Müller K, Schleicher A, Ohmes E, Ferrarini A, Kothe G (1987) Pulsed dynamic NMR of fibers from thermotropic liquid crystal polymers. Macromolecules, *20*: 2761
371. Kintanar A, Jelinski LW, Gancarz I, Koberstein JT (1986) Complex molecular motion in bulk hard segment polyurethanes: A deuterium NMR study. Macromolecules, *19*: 1876
372. Godovsky YK, Makarova NN, Petrova IM, Zhdanov AA (1984) Thermotropic mesophases in cyclolinear polymethylsiloxanes. Makromol. Chemistry, Rapid Comm. *5*: 427
373. Godovsky YK, Makarova NN, Mamaeva II (1986) A thermotropic mesophase in atactic poly[oxy(decamethylcycloxexasiloxane-2,8-diyl)]. Makromol. Chemie, Rapid Comm. *7*: 325
374. Makarova NN, Godovsky YK, Kuzmin NN (1987) The role of polymer chain tacticity of cyclolinear polydecaorganocyclohexasiloxanes in the formation of a mesomorphic state. Makromol. Chemie, *188*: 119
375. Platé NA, Shibaev VP (1974) Comb-like polymers, structure and properties. J. Polymer Sci., Macromol. Res. *8*: 117
376. Shibaev VP, Platé NA (1984) Thermotropic liquid crystalline polymers with mesogenic side groups. Adv. Polymer Sci., *60/61*: 173
377. Finkelmann H, Rehage G (1984) Liquid crystal side chain polymers. Adv. Polymer Sci. *60/61*: 99
378. Blumstein A, Hsu EC Liquid crystalline order in polymers with mesogenic side groups; in Blumstein, A. Ed.: "Liquid Crystalline Order in Polymers." Academic Press, New York, N.Y. 1978, p. 105
379. Menczel J, Wunderlich B (1981) Phase transitions in mesophase macromolecules. II. The transitions of poly(p-acryloyloxybenzoic acid). Polymer, *22*: 778
380. Pschorn U, Spiess HW, Hisgen B, Ringsdorf H (1986) Deuteron NMR study of molecular order and motion of the mesogenic side groups in liquid-crystalline polymers. Macromol. Chem. *187*: 2711 see also Spiess HW (1985) Deuteron NMR — A new tool for studying chain mobility and orientation in polymers. Adv. Polymer Sci., *66*: 23
381. Thundathvil R, Stoffer JO, Friberg SE (1980) Polymerization in lyotropic liquid crystals. I. Change of structure during polymerization. J. Poly. Sci., Polymer Chem. Ed. *18*: 2629
382. Lenz RW (1985) Balancing mesogenic and non-mesogenic groups in the design of thermotropic polyesters. Farad. Disc. Chem. Soc., *79*: 21
383. For a review see: Lagaly G (1976) Kinkblock and Gaucheblockstruckturen bimolekularer Filme. Angew. Chemie, *88*: 628 Inorganic layer compounds, phenomena of interface reactions with organic compounds Naturwissenschaften, *68*: 82 (1981)
384. Pfirrmann G, Lagaly G, Weiss A (1973) Phase transitions in complexes of nontronite with n-alkanols. Clays and Clay Minerals, *21*: 239

385. Bradley WF (1945) Molecular associations between montmorillonite and some polyfunctional organic liquids. J. Am. Chem. Soc. *67*: 975
386. MacEwan DMC (1948) Complexes of clays with organic compounds. Trans. Farad. Soc., *44*: 349
387. Lagaly G, Fitz St, Weiss A (1975) Kink block structures in clay organic complexes. Clay and Clay Minerals, *23*: 45
388. Lagaly G, Weiss A (1971) Schichteinlagerungsverbindungen als Modelle für Struktur und Strukturumwandlungen von monomolekularen und biomolekularen Schichten. I, II. Kolloid Z. Z. Polymere, *248*: 968, 979
389. Stohrer M, Noak F (1975) Magnetische Relaxationsspektroskopie an gequollenen Beidellit, einer Paraffin-Modellsubstanz. Progr. Colloid Polymer Sci., *57*: 61
390. Watanabe J, Ono H, Uematsu I, Abe A (1985) Thermotropic polypeptides. 2. Molecular packing and thermotropic behavior of poly(L-glutamate)s with low n-alkyl side chains. Macromolecules *18*: 2141
391. Uematsu I, Uematsu Y (1984) Polypeptide liquid crystals. Adv. Polymer Sci., *59*: 37
392. Herrmann-Schonherr O, Wendorff JH, Ringsdorf H, Tschirner P (1986) Structure of an aromatic polyamide with disc-like mesogens in the main chain. Makromol. Chem. Rapid Commun. *7*: 791
393. Ballauff M (1986) Rigid rod polymers having flexible side chains. I. Thermotropic poly(1,4-phenylene-2,5-dialkoxyterephthalate) s. Makromol. Chem. Rapid Commun. *7*: 407
394. Ballauff M, Schmidt GF (1987) Rigid rod polymers with flexible side chains. 2. Observations of a novel type of layered mesophase. Makromol. Chem. Rapid Commun, *8*: 93
395. Ballauff M, Schmidt GF (1987) Rigid rod polymers having flexible side chains. 3. Structural investigations on a novel layered mesophase formed by thermotropic poly(1,4-phenylene-2,5-dialkoxyterephthalate)s. Mol. Liq. Cryst. *147*: 163
396. Wunderlich B (1968) Crystallization during Polymerization. Adv. Polymer Sci., *5*: 568
397. For a recent review see, for example, Ward IM New Techniques: Developments in processing of high-performance materials. To be published, Proc. of the Second Rolduc Polymer Meeting, 1987
398. Lemstra PJ, Bastiaansen CWM, Meijer MEH (1986) Chain-extended flexible polymers. Angew. Makromol. Chemie, *145/146*: 343
399. Dobb MG, McIntyre JE (1984) Properties and application of liquid-crystalline main-chain polymers. Adv. Polymer Sci., 60/61: 61
400. Hergenrother PM (1986) High performance thermoplastics. Angew Makromol. Chemie, *145/146*: 323
401. Papkov SP (1984) Liquid crystalline order in solutions of rigid-chain polymers. Adv. Polymer Sci., *59*: 75
402. Lieser G, Schwarz G, Kricheldorf HR (1983) Polymer single crystals of poly(4-hydroxybenzoate). I. Morphology. J. Polymer Sci., Polymer Phys. Ed. *21*: 1599
403. Lieser G (1983) Polymer single crystals of poly(4-hydroxybenzoate) II. A contribution to crystal structure and polymorphism. J. Polymer Sci., Polymer Phys. Ed. *21*: 1611
404. Li L, Lieser G, Rosenau-Eichin R, Fischer EW (1987) An electron diffraction study of crystalline modifications and the liquid-crystalline structure of p-hydroxybenzoate/m-hydroxybenzoate copolymers. Makromol. Chem. Rapid Commun. *8*: 159
405. Meesiri W, Menczel J, Gaur U, Wunderlich B (1982) Phase transitions in mesophase macromolecules. III. The transitions in poly(ethylene terephthalate-*co*-p-oxybenzoate). J. Polymer Sci., Polymer Phys. Ed. *20*: 719
406. Jackson Jr WJ, Kuhfuss HF (1976) Liquid crystal polymers. I. Preparation and properties of p-hydroxybenzoic acid copolyesters. J. Polymer Sci., Polymer Chem. Ed. *14*: 2043
407. Menczel J, Wunderlich B (1980) Phase transitions in mesophase macromolecules. I. Novel behavior in the vitrification of poly(ethylene terephthalate-*co*-p-oxybenzoate). J. Polymer Sci., Polymer Phys. Ed. *18*: 1433
408. Blackwell J, Lieser G, Gutierrez GA (1983) Structure of p-hydroxybenzoate/ethylene terephthalate copolyester fibers. Macromolecules, *16*: 1418
409. Celanese Corp. (Calundann, G. W.): G.B.P. 2,006,242: US Pat. 4,161,470 (1979, priority Oct. 20, 1977 USA)
410. Cao M-Y, Wunderlich B (1985) Phase transitions in mesophase macromolecules. V. Transitions

in poly(oxy-1,4-phenylene carbonyl-*co*-oxy-2,6-naphthaloyl). J. Polymer Sci., Polymer Phys. Ed., *23*: 521

411. Butzbach GD, Wendorff JH, Zimmermann HJ (1986) Structure and structure formation of a main chain thermotropic polymer. Polymer *27*: 1337
412. Blackwell J, Biswas A, Gutierrez GA, Chivers RA (1985) X-ray analysis of the structure of liquid-crystalline copolyesters. Faraday Disc. Chem. Soc. *79*: 73 see also Chivers RA, Blackwell J, Gutierrez GA (1984) Polymer, *25*: 435
413. Windle AH, Viney C, Golombok R, Donald AM, Mitchell GR (1985) Molecular correlation in thermotropic copolyesters. Faraday Disc. Chem. Soc., *79*: 55 see also following discussion pp. 100–123
414. Bechthold H, Wendorff JH, Zimmermann HJ (1987) Studies on the nature of order in a rigid random copolymer. Makromol. Chem. *188*: 651
415. Hanna S, Windle AH (1988) Geometrical limits to order in liquid crystalline random copolymers. Polymer, *29*: 207
416. Wunderlich B (1958) Theory of cold crystallization of high polymers. J. Chem. Phys. *29*: 1395
417. Du Pont de Nemours and Co. (Kwolek, S. L.) B.P. 1, 198,081, US Pat. 3,637,606; 3,951,914 (1972, 1974, priority June 13, 1966 USA)
418. Du Pont de Nemours and Co. (Kwolek, S. L.) B.P. 1,283,064, US Pat. 3,671,542 (1972, priority June 12, 1968 USA)
419. Morgan RJ, Pruneda CO, Steele WJ (1983) The relationship between the physical structure and the microscopic deformation and failure process of poly(p-phenylene terephthalamide) fibers. J. Polymer Sci., Polymer Phys. Ed. *21*: 1757
420. Haraguchi K, Kajiyama T, Takayanagi M (1979) Effect of coagulation conditions on crystal modification of poly(p-phenylene terephthalamide). J. Appl. Polymer Sci. *23*: 915
421. Norholt MG (1974) X-ray diffraction study of poly(p-phenylene terephthalamide) fibres. Europ. Polymer J. *10*: 799 see also Norholt MG, van Aartsen JJ (1973) J. Polymer Sci., Part B., *11*: 333 and Hasegawa R, Kimoto K, Chatani Y, Tadokoro H, Meeting of the Soc. Cryst. Japan, Osaka, 1973, Abstracts pg. 21
422. Panar M, Arvakian P, Blume RC, Gardner KH, Gierke TD, Yang HH (1983) Morphology of poly(p-phenylene terephthalamide) fibers. J. Polymer Sci., Polymer Phys. Ed. *21*: 1955
423. Li L-S (1986) On the morphology and structure of poly(p-phenylene terephthalamide) single crystals and spherulites. J. Macromol. Sci.-Phys., *B25*: 215
424. English AD (1986) Solution and solid state NMR characterization of poly(p-phenylene terephthalamide). J. Polymer Sci., Polymer Phys. Ed. *24*: 805

Author Index Volumes 1–87

Subject Index

GPSR Compliance

The European Union's (EU) General Product Safety Regulation (GPSR) is a set of rules that requires consumer products to be safe and our obligations to ensure this.

If you have any concerns about our products, you can contact us on ProductSafety@springernature.com

In case Publisher is established outside the EU, the EU authorized representative is:

Springer Nature Customer Service Center GmbH
Europaplatz 3
69115 Heidelberg, Germany

Batch number: 10620881

Printed by Printforce, the Netherlands